Cambridge Elements ≡

Elements in the Philosophy of Physics
edited by
James Owen Weatherall
University of California, Irvine

The Classical–Quantum Correspondence

Benjamin H. Feintzeig
University of Washington

CAMBRIDGE
UNIVERSITY PRESS

CAMBRIDGE
UNIVERSITY PRESS

Shaftesbury Road, Cambridge CB2 8EA, United Kingdom

One Liberty Plaza, 20th Floor, New York, NY 10006, USA

477 Williamstown Road, Port Melbourne, VIC 3207, Australia

314–321, 3rd Floor, Plot 3, Splendor Forum, Jasola District Centre, New Delhi – 110025, India

103 Penang Road, #05–06/07, Visioncrest Commercial, Singapore 238467

Cambridge University Press is part of Cambridge University Press & Assessment, a department of the University of Cambridge.

We share the University's mission to contribute to society through the pursuit of education, learning and research at the highest international levels of excellence.

www.cambridge.org
Information on this title: www.cambridge.org/9781009044318
DOI: 10.1017/9781009043557

First published 2022

A catalogue record for this publication is available from the British Library.

ISBN 978-1-009-04431-8 Paperback
ISSN 2632-413X (online)
ISSN 2632-4121 (print)

The Classical–Quantum Correspondence

Elements in the Philosophy of Physics

DOI: 10.1017/9781009043557
First published online: December 2022

Benjamin H. Feintzeig
University of Washington

Author for correspondence: Benjamin H. Feintzeig, bfeintze@uw.edu

Abstract: This Element provides an entry point for philosophical engagement with quantization and the classical limit. It introduces the mathematical tools of C*-algebras as they are used to compare classical and quantum physics. It then employs those tools to investigate philosophical issues surrounding theory change in physics. It discusses examples in which quantization bears on the topics of reduction, structural continuity, analogical reasoning, and theory construction. In doing so, it demonstrates that the precise mathematical tools of algebraic quantum theory can aid philosophers of science and philosophers of physics.

Keywords: quantization, classical limit, C*-algebras, theory change, intertheoretic reduction

ISBNs: 9781009044318 (PB), 9781009043557 (OC)
ISSNs: 2632-413X (online), 2632-4121 (print)

Contents

1 Introduction

Theory change has long been a central focus for a number of issues in philosophy of science. Famously, Feyerabend and Kuhn argued that the diachronic discontinuities between scientific theories or frameworks force us to reject the fundamental core of logical empiricism.[1] Scientific realists and antirealists alike point to what has been preserved (or not) across theory change as support for their positions. And philosophical work on scientific discovery searches for prospective ways of bringing about theory change and the rationales of those methods. But while philosophers have often been interested in comparing new and old scientific theories, it is a somewhat rarer circumstance to find an entire field building precise mathematical tools for just this purpose. This is the lucky situation we find ourselves in for comparing classical and quantum physics, where the tools of *quantization theory* are available for analyzing the classical–quantum correspondence. The aim of this contribution is to provide an introduction to those mathematical tools and to demonstrate how they can bear on topics of philosophical interest. Such work should be of interest to those who are primarily interested in the philosophical and interpretive issues that arise in quantum theories, as well as for those who want to use the classical–quantum transition as a case study for other general issues in philosophy of science.

1.1 Motivations

Philosophical work on the classical–quantum correspondence aims to help us understand the relationship between new quantum theories and their old classical predecessors from which they are constructed. Part of the motivation for such work comes from a number of outstanding issues *within* modern physics.

While many theoretical physicists are working toward applying the framework of quantum theory to an understanding of gravitational phenomena, there is a wide array of different approaches to constructing a theory of quantum gravity.[2] This proliferation of methodologies suggests disagreement about even what we are aiming for, or what it might mean to construct a quantum analog of our current best gravitational physics.

Indeed, a similar issue arises before even entering the contentious arena of quantum gravity, where we do not yet have the ability to make empirically testable predictions from any new theories. Even the standard model of particle

---

[1] See, for example, Kuhn (1962, 1984) and Feyerabend (1962).
[2] See, for example, Callender and Huggett (2001) or Crowther (2021).

physics, which is now extremely well confirmed by experiments, lacks a firm mathematical foundation.[3] The standard model is based on the quantum theory of interacting fields, where roughly material particles are represented by excitations in fermionic fields and force-carrying particles correspond to excitations in bosonic fields. The form of the interactions between these different particles, or the mutual actions of matter fields on force fields, is governed by modern gauge theory or Yang–Mills theory. However, the mathematical foundation for quantum Yang–Mills theory is currently unknown, and it is an open problem to reproduce even some of the basic properties that physicists rely on in their heuristic formulations to arrive at the predictions of the standard model. There are again a variety of different approaches to constructing quantum field theories, none of which has successfully led to a full understanding of interacting particles. The very process of theory construction is at issue in these cases.

Why should the methods for constructing quantum theories be of interest to philosophers? It might (naïvely) be thought that what matters philosophically is not how one arrives at a new theory – the idea could have come in a dream, or as a lucky guess – but rather how one rationally supports or confirms that theory with evidence. However, in these open areas we see working mathematical and theoretical physicists at least attempting to justify their methods for theory construction and provide rationales for their approaches. These purported justifications and rationales of course rely on an understanding of what has worked in the past, in both the classical theories we seek to quantize and in how the procedures to construct existing quantum theories have been successful. This opens up a host of deep philosophical issues concerning intertheory relations that, speculatively, have the potential to make an impact on the forefront of the development of physics.

Much more philosophical work on quantization is required to seriously engage with the as yet unknown foundations of the quantum theories just mentioned. The topics covered in this Element will bear on those issues only tangentially. Our task will primarily be to analyze the classical–quantum correspondence in much simpler and better understood cases. But with these motivations from current and aspirational physics, we hope that this Element will serve as an invitation for other researchers to push philosophical work on quantization forward.

---

[3] See Jaffe and Witten (2000) and Douglas (2004). Specifically, we do not yet have a nonperturbative formulation of the standard model in a continuous spacetime.

1.2 History

Correspondence between old and new physics was brought to the fore during the development of quantum mechanics, especially in the work and influence of Niels Bohr between 1913 and 1930. Bohr's so-called "correspondence principle" underwent considerable evolution during this time period.[4]

In the old quantum theory, the correspondence principle played a role in fixing features of semiclassical atomic models. Famously, Bohr's 1913 trilogy proposed a model of simple atoms containing a discrete set of possible circular electron orbits, thought of as classical trajectories of particles around the nucleus.[5] The energies of electrons in such orbits could then only take values allowed by a quantization condition analogous to those employed by Planck in his investigation of black-body radiation[6] and Einstein in his light quantum theory.[7] Bohr realized that transitions of electrons between such orbits of fixed energy, which became known as "quantum jumps," could produce radiation with characteristic frequencies corresponding to the differences of those energy levels, and thus reproduce the phenomenology of spectral lines. While such a model is explicitly nonclassical in its description of both mechanical motion of subatomic particles and the production of electromagnetic radiation, Bohr required that in the limit of large quantum numbers (higher values for the electron energy levels), the radiation produced should match that of classical electromagnetic theory. This version of the correspondence principle was used, for example, to calculate the predicted intensities of spectral lines in atomic models.

The atomic models of the old quantum theory produced notable successes in the period 1913–1923,[8] especially in the hands of Sommerfeld in Munich along with Epstein, Pauli, and Heisenberg. For example, generalization to elliptical orbits produced a multiplicity of quantum numbers, which in turn allowed the theory to account for aspects of the Zeeman and Stark effects. Ultimately, however, the old quantum theory faced trouble in accounting for more complicated spectra, including that of helium. By the early 1920s, Bohr, Pauli, and Born saw these problems as a crisis for the quantum theory.

---

[4] There are many historical sources for the history of quantum mechanics, and our discussion and references in this section will only provide a small sample. For general historical background, see, for example, Jammer (1966) and Forman (1971). For another reference tying together the history and philosophy of the classical–quantum correspondence, see Bokulich (2008).

[5] See Duncan and Janssen (2019, Ch. 4) or Darrigol (1992, Ch. V).

[6] See, for example, Kuhn (1984, 1987) or Duncan and Janssen (2019, Ch. 2).

[7] See, for example, Duncan and Janssen (2019, Ch. 3).

[8] See Duncan and Janssen (2019, Ch. 5–7) or Darrigol (1992, C. VI–VIII).

The correspondence principle began to take a new form as theorists responded to the crisis generated by the problems of the old quantum theory. Bohr turned his attention to understanding the production of radiation as he, along with Kramers and Slater, proposed a theory of radiation in terms of virtual oscillators,[9] which resulted in such radical predictions as violation of the conservation of energy. While scattering experiments upheld energy conservation and thus provided evidence against the Bohr–Kramers–Slater theory, attention toward radiation ultimately provided fruitful ground for the application of quantum ideas. Kramers and Heisenberg turned toward investigating dispersion[10] and applied the correspondence principle to develop formal analogies between classical and quantum physics. Their strategy involved replacing continuous classical quantities with discrete quantum counterparts, a development which provided formulas that by 1924–25 paved the way for Heisenberg's reinterpretation.

While Heisenberg presented his ideas as echoing positivist credos to use only observable quantities, his work developing matrix mechanics built strongly upon the theoretical foundation and formalism of classical physics, especially in the central role he gave to the commutation relations. In Göttingen, along with Born and Jordan, this use of the correspondence principle via formal analogies gave rise to the full mathematical apparatus of matrix mechanics.[11] The abstract matrix theory faced competition, however, from Schrödinger's very different classical viewpoint in his wave mechanics, which built upon ideas of Einstein and de Broglie. Schrödinger's theory had a positive reception – because the tools of wave mechanics were visualizable and familiar to many physicists – and in response, the matrix theorists Born, Heisenberg, Jordan, and Pauli redoubled their efforts.[12] Eventually, this drove the need for an interpretation of the matrix formalism that had been built upon formal analogies to classical physics, which Heisenberg provided through his uncertainty relation, connecting the algebraic relations directly to limitations on physical experiments.[13] On the other hand, Born's probabilistic interpretation of the wavefunction filled in Schrödinger's picture so that there were two formalisms, known to be equivalent in their descriptions of at least some systems, yet providing seemingly different physical pictures.[14] At the same time, Dirac's work on analogies between the formalisms of classical and quantum mechanics

---

[9] See Darrigol (1992, Ch. IX) or Beller (1999, Ch. 2).
[10] See Darrigol (1992, Ch. IX–X), Duncan and Janssen (2007a,b, 2022).
[11] See Darrigol (1992, Ch. X) and Beller (1999, Ch. 2–3).
[12] See Darrigol (1992, Ch. XIII) and Beller (1999, Ch. 2–3).
[13] See Beller (1999, Ch. 4–5).
[14] See Beller (1999, Ch. 2).

drew attention to the correspondence between classical Poisson brackets and commutators and led to his theory of q-numbers.[15] The further work of Dirac and Jordan developed transformation theory, which provided a way of unifying the matrix and wave theories.[16]

There are two strands remaining for our purposes at the end of this exciting period for the quantum theory during the 1920s. First, one of Bohr's central interpretive contributions to the new quantum theory[17] (setting aside the principle of complementarity, which will play no role in this Element) was to add to the correspondence principle his doctrine of classical concepts. For Bohr, this played a crucial role in his analysis of measurement setups in his disagreements with Einstein,[18] which led to the consolidation of the Copenhagen view. For us, it suffices to note that Bohr set a precedent for taking the formal analogies between classical and quantum mechanics that motivated the construction of the new theory and giving those same analogies a substantive place in the interpretation of the new physics.

Second, the important mathematical developments among Hilbert's circle in Göttingen, which paralleled and followed the development of quantum theory by physicists, will play an important role in our discussion to follow. It is well known that von Neumann provided the full mathematical formulation of the quantum theory in terms of the abstract definition of Hilbert spaces, so that the decade was capped by two beautiful accounts of the new quantum theory from Dirac (1930) and von Neumann (1932). von Neumann's subsequent work on operator theory and operator algebras (Murray and von Neumann, 1936) similarly laid the foundation for much of the mathematical framework we will use in what follows. But of equal importance for us is the legacy of Weyl (1950) and Wigner (1959), whose work brought considerations of symmetry into quantum mechanics. In the work of Groenewold (1946) and Moyal (1949), this led to the phase space formulation of quantum mechanics, and later to the advent of contemporary quantization theory.

What follows is an attempt to continue in this tradition by exploring interpretive and philosophical issues through mathematical comparisons of classical and quantum theories. The all too brief historical narrative just presented is intended only as a backdrop for the remainder of this Element, and we make no attempt to tie the discussion presented here to the views of any of the thinkers just mentioned. We refer the reader to the references for a more detailed and complete

---

[15] See Darrigol (1992, Ch. XI–XIII).

[16] See Darrigol (1992, Ch. XIII) and Duncan and Janssen (2012).

[17] See Beller (1999, 6–9, 11–12) and Howard (1986).

[18] See Beller (1999, Ch. 7) and Fine (1988).

history than can be compressed in this short section. One central goal of this Element is to ensure that the history does not end here, but rather is continued by those who read the pages that follow.

1.3 Required Background

There is no way to provide a self-contained and comprehensive introduction to the issues we will grapple with here. So while this Element is intended as an (opinionated) entry point to the literature for researchers or graduate students in the mathematical and philosophical foundations of physics, we must still assume some background.

On the philosophical side, we will assume some familiarity with central issues in general philosophy of science. Readers who have surveyed the topics in Curd et al. (2013) will be well prepared. We will provide short descriptions of general philosophical questions as we come to them, while providing references in the main text. However, we make no suggestion that our reference list is complete or provides full historical context.

In terms of physics, we presume knowledge of the standard formulation of quantum mechanics. We will at times deal with systems with symmetries as examples, but the majority of our discussion focuses on even just the kinematics of simple quantum mechanical point particles. We assume a working knowledge of material at the level of Sakurai (1994). We will also engage with free quantum field theories, primarily through the example of a scalar field. It will be helpful for readers to have some prior familiarity with the standard (Fock space) formulation of free quantum field theory as found, for example, in Peskin and Schroeder (1995).

The mathematical background for this Element is more demanding. We will assume that readers have background in topology and operator theory on infinite dimensional Hilbert spaces, roughly equivalent to one year of graduate coursework in functional analysis at the level of Reed and Simon (1975). This will allow us to jump directly into the more abstract framework of C*-algebras, which we will use to unify classical and quantum mechanics in the formulation of both quantization and the classical limit. We will devote a full section (Section 2) to an introduction to the mathematics of C*-algebras for unfamiliar readers; this material will be best understood by readers who already have experience in functional analysis. While not all of even the standard topics in functional analysis will be relevant to our discussion, we believe that the reader should have a solid foundation in the topic to put C*-algebras in context. Finally, but less importantly, we mention that the theory of topological groups will appear in some of our examples, for which one can use Rudin (1962) as a reference. Similarly, some of the more technical examples involve Lie groups and associated structures. We will not cover the background for Lie groups, and these examples can be skipped without loss of continuity.

In Section 2, we will endeavor to provide a brief introduction for non-experts to the fundamentals of the mathematics required for analyzing the relationship between classical and quantum mechanics. Our introduction will assume and build upon the expected background just described. We hope to provide the level of proficiency required for understanding the discussion that follows in the remainder of the Element, as well as for basic comprehension of the arguments for key technical results.

1.4 Outline

This Element is split into two parts. Sections 2 and 3 provide the technical background. Sections 4, 5, and 6 use those tools to discuss a number of philosophical issues.

The technical background is organized as follows. Section 2 provides an introduction to the mathematical theory of C*-algebras that will be used throughout this Element. We focus on illustrations through examples in both classical and quantum physics. Section 3 then uses this mathematical framework to formulate the theory of quantization, which provides the tools for analyzing the classical–quantum correspondence. We illustrate these tools in application to examples, as well as provide a brief description of the relationship between the mathematical tools we employ in this Element and a number of alternative approaches.

The remainder of the Element surveys interconnected philosophical topics. Section 4 analyzes whether the $\hbar \to 0$ limit, as formulated through the tools of quantization, can be used to provide a reductive explanation of classical physics on the basis of quantum physics. Section 5 considers ways in which the tools of quantization and the classical limit can aid the interpretation of quantum theories, including its relation to structural realism and analogical reasoning. Section 6 then directly tackles the use of quantization in the construction of new quantum theories, and the possibility of providing a rationale for certain methods of construction. Section 7 provides a brief conclusion.

2 C*-Algebras of Physical Quantities

The purpose of this section and the next is to familiarize the reader with the modern mathematical tools for formulating the $\hbar \to 0$ limit, as a first step toward analyzing the classical–quantum correspondence. This section provides background on the general theory of C*-algebras, which serves as a framework for the formulation of the $\hbar \to 0$ limit. Next, Section 3 describes the formulation of the $\hbar \to 0$ limit in terms of quantization maps, as well as the associated mathematical structures. Both sections strive to help readers new to the subject by providing examples that draw on other familiar mathematical concepts.

2.1 Basic Theory of C*-Algebras

Our starting point for comparing classical and quantum physics will be to put both theories in a common mathematical language. We will do this by analyzing the algebraic structure of the physical quantities represented by both theories. One can represent (bounded) physical quantities in both classical and quantum theories by *C*-algebras*.[19]

Definition 1 A *C*-algebra* $\mathfrak{A}$ is a set of elements with the following structures:

- *addition*: a binary operation $\mathfrak{A} \times \mathfrak{A} \to \mathfrak{A}$ denoted by $(A, B) \mapsto A + B$
- *scalar multiplication*: an operation $\mathbb{C} \times \mathfrak{A} \to \mathfrak{A}$ denoted by $(\alpha, A) \mapsto \alpha \cdot A$
- *multiplication*: a binary operation $\mathfrak{A} \times \mathfrak{A} \to \mathfrak{A}$ denoted by $(A, B) \mapsto A \cdot B$
- *involution*: a unary operation $\mathfrak{A} \to \mathfrak{A}$ denoted by $A \mapsto A^*$
- *norm*: a function $\mathfrak{A} \to \mathbb{R}$ denoted by $A \mapsto \|A\|$

Together, these structures satisfy the following conditions:

(i) $\mathfrak{A}$ is a *complex normed vector space* under addition and scalar multiplication.[20]

(ii) $\mathfrak{A}$ is an *algebra*, that is, multiplication is associative and distributive over addition: for any $A, B, C \in \mathfrak{A}$,

$$A \cdot (B \cdot C) = (A \cdot B) \cdot C$$
$$A \cdot (B + C) = A \cdot B + A \cdot C. \tag{2.1}$$

---

[19] For further mathematical introduction to C*-algebras, see Kadison and Ringrose (1997) or Sakai (1971), Bratteli and Robinson (1987, 1996), and the appendix to Landsman (2017). For an introduction to the application of C*-algebras in quantum physics, see again Bratteli and Robinson (1987, 1996) and Landsman (2017) as well as Haag (1992) and Emch (1972). For introductions intended for a philosophical audience, see Halvorson (2007) and Ruetsche (2011).

[20] This means that $\mathfrak{A}$ forms an abelian group under addition and a $\mathbb{C}$-module under scalar multiplication. Further, the norm is positive definite, absolutely homogeneous, and satisfies the triangle inequality.

(iii) $\mathfrak{A}$ is *involutive*: for any $A, B \in \mathfrak{A}$, and $\alpha \in \mathbb{C}$,

$$(A + B)^* = A^* + B^*$$
$$(\alpha A)^* = \overline{\alpha} A^*$$
$$(AB)^* = B^* A^* \tag{2.2}$$
$$(A^*)^* = A.$$

(iv) $\mathfrak{A}$ is a *Banach space*: for any sequence $\{A_n\}_{n \in \mathbb{N}}$ that is *Cauchy* in the sense that, as $m, n \to \infty$,

$$\|A_n - A_m\| \to 0, \tag{2.3}$$

there is an element $A \in \mathfrak{A}$ such that A_n converges to A in norm, that is,

$$\|A_n - A\| \to 0. \tag{2.4}$$

(v) $\mathfrak{A}$ is a *Banach algebra*, that is, multiplication is jointly continuous with respect to the norm: for any $A, B \in \mathfrak{A}$,

$$\|AB\| \leq \|A\| \|B\|. \tag{2.5}$$

(vi) $\mathfrak{A}$ satisfies the *C*-identity*: for any $A \in \mathfrak{A}$,

$$\|A^* A\| = \|A\|^2. \tag{2.6}$$

Note that the multiplication operation in a C*-algebra may be either *commutative* – in the sense that $AB = BA$ for all $A, B \in \mathfrak{A}$ – or *noncommutative*. This freedom in the definition allows one to use C*-algebras to represent physical quantities in both classical and quantum physics. Two initial examples will illustrate their use.

Example 1 (*n* classical particles on a line) *For a system of n classical particles moving in* $\mathbb{R}$, *the physical quantities are represented by functions on the phase space of the system. The phase space* $\mathbb{R}^{2n}$ *consists in the possible values of position and momentum for each particle. A physical quantity assigns a (possibly complex, for generality) numerical value to each point in the phase space. Consider the collection of continuous complex-valued functions on the phase space vanishing at infinity, denoted by* $C_0(\mathbb{R}^{2n})$.[21] *Define pointwise algebraic operations for each* $f, g \in C_0(\mathbb{R}^{2n})$, $\alpha \in \mathbb{C}$, *and* $x \in \mathbb{R}^{2n}$ *by*

---

[21] A complex-valued function f on a topological space X *vanishes at infinity* if for every $\epsilon > 0$, the set $\{x \in X \mid |f(x)| \geq \epsilon\}$ is compact. If X itself is compact, then $C_0(X)$ is the same as the collection $C(X)$ of all continuous functions on X.

$$(f + g)(x) := f(x) + g(x)$$

$$(\alpha f)(x) := \alpha \cdot f(x)$$

$$(fg)(x) := f(x)g(x) \tag{2.7}$$

$$(f^*)(x) := \overline{f(x)}.$$

Endow $C_0(\mathbb{R}^{2n})$ with the supremum norm[22]

$$\|f\| := \sup_{x \in \mathbb{R}^{2n}} |f(x)|. \tag{2.8}$$

Then it follows that $C_0(\mathbb{R}^{2n})$ is a commutative C*-algebra of physical quantities because, for example, adding and multiplying elements of $C_0(\mathbb{R}^{2n})$ always yields an element of $C_0(\mathbb{R}^{2n})$.[23]

Example 2 (*n* quantum particles on a line) *For a system of n quantum particles moving in $\mathbb{R}$, the physical quantities are represented by linear operators on the Hilbert space of the system. The Hilbert space $L^2(\mathbb{R}^n)$ consists in the possible wavefunctions for the joint system. A physical quantity is a linear operator that assigns a (possibly complex, for generality) expectation value to each wavefunction via the Born rule. Consider the collection of compact linear operators on the Hilbert space, denoted by $\mathcal{K}(L^2(\mathbb{R}^n))$.*[24] *Define pointwise algebraic operations for each $A, B \in \mathcal{K}(L^2(\mathbb{R}^n))$, $\alpha \in \mathbb{C}$, and $\psi \in L^2(\mathbb{R}^n)$ by*

$$(A + B)\psi := A\psi + B\psi$$

$$(\alpha A)\psi := \alpha \cdot A\psi$$

$$(AB)\psi := A(B\psi) \tag{2.9}$$

$$(A^*)\psi := A^\dagger \psi,$$

where $A^\dagger$ is the Hermitian adjoint of A. Endow $\mathcal{K}(L^2(\mathbb{R}^n))$ with the operator norm[25]

$$\|A\| := \sup_{\psi \in L^2(\mathbb{R}^n)} \frac{\|A\psi\|}{\|\psi\|}. \tag{2.10}$$

---

[22] Every continuous function *f* vanishing at infinity is *bounded* in the sense that $\|f\|$ exists and is finite.

[23] For more on commutative C*-algebras, see §4.4 of Kadison and Ringrose (1997) or Gamelin (1969). For their role in classical physics, see chapter I of Landsman (1998a).

[24] A linear operator A on a Hilbert space $\mathcal{H}$ is called *compact* just in case for every bounded sequence $\{\psi_n\}_{n \in \mathbb{N}}$ in $\mathcal{H}$, the sequence $\{A\psi_n\}$ contains a converging subsequence.

[25] Every compact operator A is *bounded* in the sense that $\|A\|$ exists and is finite.

Then it follows that $\mathcal{K}(L^2(\mathbb{R}^n))$ is a noncommutative C*-algebra of physical quantities because, for example, adding and multiplying elements of $\mathcal{K}(L^2(\mathbb{R}^n))$ always yields an element of $\mathcal{K}(L^2(\mathbb{R}^n))$.[26]

These examples demonstrate that C*-algebras of physical quantities appear in both classical and quantum physics. Commutative C*-algebras of functions on a phase space represent collections of physical quantities in classical physics, while noncommutative C*-algebras of linear operators on a Hilbert space represent collections of physical quantities in quantum physics. However, showing that C*-algebras appear in the mathematical formalism of each of these theories is not enough for our purposes. Our task in the next section is to show that the structure of a C*-algebra of physical quantities captures enough of the physically relevant information encoded in a theory to be an apt tool for comparing classical and quantum physics.

To show how the structure of a C*-algebra captures physically significant information, we will in the next section review how the abstract algebraic structure of a C*-algebra suffices to determine the probabilistic structure of a space of physical states. Then we will review the sense in which the preceding examples of C*-algebras are representative: every commutative C*-algebra can be understood as an algebra of functions on a topological space, and every noncommutative C*-algebra can be understood as an algebra of linear operators on a Hilbert space. This, in turn, demonstrates that the topological structure of a classical phase space and the vector space structure of a quantum Hilbert space are determined by the structure of a C*-algebra of physical quantities.

2.2 States

We begin by defining the state space of a C*-algebra.

Definition 2 Given a C*-algebra $\mathfrak{A}$, a state on $\mathfrak{A}$ is a linear mapping $\omega : \mathfrak{A} \to \mathbb{C}$ satisfying the following conditions:

(i) ω is *bounded*: there is a constant $C \in \mathbb{R}$ such that $|\omega(A)| \leq C\|A\|$ for each $A \in \mathfrak{A}$. The constant $\|\omega\| := \sup_{A \in \mathfrak{A}} \frac{|\omega(A)|}{\|A\|}$ is called the *norm* of ω, so a bounded functional is one with finite norm.

---

[26] For more mathematical details, see Reed and Simon (1975) – in particular, chapter II for Hilbert spaces, chapter VI for general operator theory, and chapter VI.5–6 for compact operators and associated structures.

(ii) ω is *positive*: for each $A \in \mathfrak{A}$, $\omega(A^*A) \geq 0$. Algebra elements of the form A^*A are also called *positive*, so a positive functional is one that assigns positive elements to nonnegative numbers.

(iii) ω is *normalized*: $\|\omega\| = 1$. If the algebra $\mathfrak{A}$ contains a multiplicative identity I satisfying $AI = IA = A$ for all $A \in \mathfrak{A}$, then (given that ω is positive) this is equivalent to the condition $\omega(I) = 1$.

The *state space* $S(\mathfrak{A})$ is the collection of all states on $\mathfrak{A}$.[27]

Each state $\omega \in S(\mathfrak{A})$ is a possible assignment of expectation values to all of the physical quantities in $\mathfrak{A}$. One can check that the state space is a *convex* subset of the dual space $\mathfrak{A}^*$ (consisting of all bounded linear functionals on $\mathfrak{A}$) – that is, for each pair of states $\omega_1, \omega_2 \in S(\mathfrak{A})$ and real numbers $a_1, a_2 > 0$ with $a_1 + a_2 = 1$, the functional $\omega \colon \mathfrak{A} \to \mathbb{C}$ defined by

$$\omega(A) = a_1\omega_1(A) + a_2\omega_2(A) \tag{2.11}$$

for all $A \in \mathfrak{A}$ is a state. A state ω is called a *mixed state* if there exist distinct states $\omega_1 \neq \omega_2$ such that Eq. (2.11) holds. Otherwise, ω is called a *pure state*. Pure states are extreme points in the convex space $S(\mathfrak{A})$, and the collection of all pure states on $\mathfrak{A}$ is denoted $\mathcal{P}(\mathfrak{A})$.

To illustrate the notion of a state, we consider states on each of our examples of C*-algebras: $C_0(\mathbb{R}^{2n})$ and $\mathcal{K}(L^2(\mathbb{R}^n))$. First, we consider the classical case.

Example 3 (probability measures on $\mathbb{R}^{2n}$) *We consider states on $C_0(\mathbb{R}^{2n})$. Recall that a probability measure on $\mathbb{R}^{2n}$ is a function $\mu \colon \Sigma \to \mathbb{R}$ on a σ-algebra[28] Σ of subsets of $\mathbb{R}^{2n}$ satisfying:*

(i) $\mu(X) \geq 0$ for every $X \in \Sigma$;

(ii) $\mu(\mathbb{R}^{2n}) = 1$; and

(iii) whenever $X_i \in \Sigma$ is a countable sequence of pairwise disjoint measurable subsets (i.e., $X_i \cap X_j = \varnothing$ for distinct $i \neq j$), then

$$\mu\left(\bigcup_i X_i\right) = \sum_i \mu(X_i). \tag{2.12}$$

---

[27] For an introduction to states, see chapter 4.3 of Kadison and Ringrose (1997). For more on the geometry of state spaces, see Alfsen and Shultz (2001).

[28] A σ-algebra for a set X is a collection of subsets of X that is closed under complementation and countable unions. We assume throughout that for measures on topological spaces, the corresponding σ-algebra contains all of the open sets.

Given a probability measure μ on $\mathbb{R}^{2n}$ (satisfying a technical condition of regularity),[29] define $\omega_\mu : C_0(\mathbb{R}^{2n}) \to \mathbb{C}$ by

$$\omega_\mu(f) := \int_{\mathbb{R}^{2n}} f \, d\mu \tag{2.13}$$

for each $f \in C_0(\mathbb{R}^{2n})$. Then ω_μ is a state on $C_0(\mathbb{R}^{2n})$.

The positivity of ω_μ follows from condition (i), the boundedness and normalization of ω_μ follow from condition (ii), and the linearity of ω_μ follows from condition (iii), which implies the linearity of the integral.

Further, consider a delta function measure μ_x for a fixed choice of $x \in \mathbb{R}^{2n}$ defined by

$$\mu_x(X) := \begin{cases} 1 & \text{if } x \in X \\ 0 & \text{otherwise} \end{cases} \tag{2.14}$$

for all $X \in \Sigma$. Then μ_x is a probability measure and one can show that ω_{μ_x} is a pure state. The pure state ω_{μ_x} reproduces all of the (determinate) expectation values of the physical quantities at the point x,[30] that is, for any $f \in C_0(\mathbb{R}^{2n})$,

$$\omega_{\mu_x}(f) = f(x). \tag{2.15}$$

Probability measures on $\mathbb{R}^{2n}$ determine states on $C_0(\mathbb{R}^{2n})$ via their expectation value assignments. The converse is also true, namely that, in general, a state on $C_0(\mathbb{R}^{2n})$ determines a probability measure on $\mathbb{R}^{2n}$ matching its expectation value assignments, as captured by the following theorem.

Theorem 1 (Riesz–Markov) *For any state ω on $C_0(\mathbb{R}^{2n})$, there is a probability measure μ on $\mathbb{R}^{2n}$ such that $\omega = \omega_\mu$, with ω_μ as defined in Eq. (2.13).*

Next, we turn to states on a quantum algebra.

Example 4 (density operators on $L^2(\mathbb{R}^n)$) *We consider states on $\mathcal{K}(L^2(\mathbb{R}^n))$. Recall that a density operator on $L^2(\mathbb{R}^n)$ is a bounded, trace class,[31] linear operator ρ on $L^2(\mathbb{R}^n)$ satisfying the following conditions:*

---

[29] Specifically, the measure must be both inner and outer regular in the sense that for any $A \in \Sigma$, (i) $\mu(A)$ is the supremum of $\mu(K)$ over compact subsets $K \subseteq A$, and (ii) $\mu(A)$ is the infimum of $\mu(U)$ over open sets U such that $A \subseteq U$.

[30] See also chapter IV.304 of Reed and Simon (1975), or chapter B.5 of Landsman (2017).

[31] The *trace* of an operator A on a Hilbert space $\mathcal{H}$ is $Tr(A) := \sum_k \langle \psi_k, A\psi_k \rangle$, where $\{\psi_k\}$ is any orthonormal basis for $\mathcal{H}$. This expression is not, in general, well-defined because it may fail to converge or fail to be unique, so we must restrict the domain of Tr to the so-called trace class operators. A bounded operator A is called *trace class* if $Tr((A^*A)^{1/2})$ exists (i.e., the sum converges) and is finite, in which case it is independent of the chosen basis. The trace class operators form an ideal in the compact operators, which means that if ρ is a trace class operator and A is a compact operator, then ρA and $A\rho$ are trace class.

(i) for each $\psi \in L^2(\mathbb{R}^n)$, $\langle \psi, \rho\psi \rangle \geq 0$; and
(ii) $Tr(\rho) = 1$.

Given a density operator ρ on $L^2(\mathbb{R}^n)$, define $\omega_\rho : \mathcal{K}(L^2(\mathbb{R}^n)) \to \mathbb{C}$ by

$$\omega_\rho(A) := Tr(\rho A) \tag{2.16}$$

for each $A \in \mathcal{K}(L^2(\mathbb{R}^n))$. Then ω_ρ is a state on $\mathcal{K}(L^2(\mathbb{R}^n))$.

The positivity of ω_ρ follows from condition (i), the boundedness and normalization of ω_ρ follow from condition (ii), and the linearity of ω_ρ follows from the linearity of the trace.

Further, consider a rank one projection P_ψ for a unit vector $\psi \in L^2(\mathbb{R}^n)$ defined by

$$P_\psi \varphi := \langle \varphi, \psi \rangle \psi \tag{2.17}$$

for all $\varphi \in L^2(\mathbb{R}^n)$. Then P_ψ is a density operator and ω_{P_ψ} is a pure state. The pure state ω_{P_ψ} reproduces all the expectation values determined by the Born rule for the vector ψ,[32] that is, for any $A \in \mathcal{K}(L^2(\mathbb{R}^n))$,

$$\omega_{P_\psi}(A) = \langle \psi, A\psi \rangle. \tag{2.18}$$

Density operators on $L^2(\mathbb{R}^n)$ determine states on $\mathcal{K}(L^2(\mathbb{R}^n))$ via their expectation value assignments. Again, the converse is also true that, in general, a state on $\mathcal{K}(L^2(\mathbb{R}^n))$ determines a density operator matching its expectation value assignments.

Theorem 2 *For any state ω on $\mathcal{K}(L^2(\mathbb{R}^n))$, there is a density operator ρ on $L^2(\mathbb{R}^n)$ such that $\omega = \omega_\rho$, with ω_ρ as defined in Eq. (2.16).*

The foregoing shows that states on C*-algebras correspond with more familiar notions of physical states – probability measures on classical phase spaces and density operators on quantum Hilbert spaces. Hence, one can interpret both classical and quantum states as instances of expectation value assignments in a single formalism.

2.3 Representations

This section reviews two fundamental theorems concerning the structure of C*-algebras. The first demonstrates that every commutative C*-algebra can be understood as an algebra of functions on some topological space. The second

---

[32] See also chapter VI.6 of Reed and Simon (1975) or chapter B.9 of Landsman (2017).

demonstrates that every non-commutative C*-algebra can be understood as an algebra of operators on a Hilbert space. In both cases, the strategy is to provide a structure-preserving map from the C*-algebra we start with to an algebra of either functions or operators. We make precise what is meant by a structure-preserving map as follows.

Definition 3 A *-*homomorphism* between C*-algebras $\mathfrak{A}$ and $\mathfrak{B}$ is a map $\varphi \colon \mathfrak{A} \to \mathfrak{B}$ such that for all $A, B \in \mathfrak{A}$, and $\alpha \in \mathbb{C}$

(i) $\varphi(A + B) = \varphi(A) + \varphi(B)$;
(ii) $\varphi(\alpha A) = \alpha \cdot \varphi(A)$;
(iii) $\varphi(AB) = \varphi(A)\varphi(B)$; and
(iv) $\varphi(A^*) = \varphi(A)^*$.

These conditions further imply that φ is continuous in the sense that $\|\varphi(A)\| \leq \|A\|$.[33] If a *-homomorphism is a bijection, then it is called a *-*isomorphism*.

We can illustrate the notion of a structure-preserving map with the following construction, which we will use later on.

Example 5 (quotients of C*-algebras) *Suppose $\mathfrak{A}$ is a C*-algebra and K is a closed two-sided ideal in $\mathfrak{A}$, that is, a linear subspace $K \subseteq \mathfrak{A}$ such that whenever $A \in K$ and $C \in \mathfrak{A}$, we have $A^*, AC, CA \in K$. Define the quotient C*-algebra $\mathfrak{A}/K$ as the collection of equivalence classes $[A]$ for $A \in \mathfrak{A}$ under the equivalence relation*

$$A \sim B \qquad iff \qquad A = B + C \text{ for some } C \in K. \qquad (2.19)$$

$\mathfrak{A}/K$ *is a C*-algebra with operations*

$$[A] + [B] := [A + B]$$

$$\alpha[A] := [\alpha A]$$

$$[A][B] := [AB] \qquad\qquad (2.20)$$

$$[A]^* := [A^*]$$

$$\|[A]\| := \inf_{C \in K} \|A + C\|$$

---

[33] This follows from a central implication of the C*-identity for the norm in a C*-algebra: the norm of certain elements can be computed from their spectral radius, which is determined completely from the algebraic structure. See, for example, Prop. 4.1.1 on p. 238 of Kadison and Ringrose (1997).

for all $A, B \in \mathfrak{A}$. Consider the canonical quotient map $\varphi \colon \mathfrak{A} \to \mathfrak{A}/K$ defined by

$$\varphi(A) := [A] \tag{2.21}$$

for all $A \in \mathfrak{A}$. Often the equivalence class $[A] \in \mathfrak{A}/K$ for $A \in \mathfrak{A}$ will be denoted by $(A + K) \in \mathfrak{A}/K$. It follows that φ is a surjective *-homomorphism.[34]

We now state the fundamental theorem for commutative C*-algebras and sketch a proof.

Theorem 3 (Commutative Gelfand–Naimark) *If $\mathfrak{A}$ is a commutative C*-algebra, then $\mathfrak{A}$ is *-isomorphic to $C_0(X)$ for some locally compact Hausdorff space X.*

In particular, if $\mathfrak{A}$ is unital *in the sense that it contains a multiplicative identity I satisfying $AI = IA = A$ for each $A \in \mathfrak{A}$, then $\mathfrak{A}$ is *-isomorphic to the collection $C(X)$ of all continuous functions on some compact Hausdorff space X.*[35]

The basic strategy of the proof is to construct X as the pure state space of the algebra itself, and have each algebra element $A \in \mathfrak{A}$ assign to each pure state $\omega \in \mathcal{P}(\mathfrak{A})$ the expectation value that ω would assign to A. To endow $\mathcal{P}(\mathfrak{A})$ with a topology, we consider it as a subspace of the dual space $\mathfrak{A}^*$ and endow $\mathfrak{A}^*$ with the weak* topology. We define the weak* topology in general for the dual to an arbitrary Banach space, since we will have the opportunity to use it later on.

Definition 4 For a Banach space V, the *weak* topology* on V^* (also denoted the $\sigma(V^*, V)$-*topology*) is characterized by the following condition for convergence. A net $\{\omega_\lambda\}$ converges to ω in V^* just in case

$$\text{for every } v \in V, \omega_\lambda(v) \to \omega(v). \tag{2.22}$$[36]

Proof sketch of Theorem 3: We consider only the simpler case where $\mathfrak{A}$ is unital. First, one can establish that if ω is a state on $\mathfrak{A}$, then ω is pure iff $\omega(AB) = \omega(A)\omega(B)$ for every $A, B \in \mathfrak{A}$. It follows that the pure state space is a closed

---

[34] See also Dixmier (1977) or Feintzeig (2018a) for more details.

[35] See chapter 4.4 of Kadison and Ringrose (1997) or chapter C.2 of Landsman (2017).

[36] A *net* is a generalization of a sequence where the indexing set can be an arbitrary directed set (including, but not necessarily, $\mathbb{N}$). When a topology is second countable, it can be characterized just by conditions of convergence for sequences, but if – as in many topologies of interest in the study of operator algebras – it is not second countable, then one must work more generally with nets.

subset of the unit ball of $\mathfrak{A}^*$ in the weak* topology. But the Banach–Alaoglu theorem (Reed and Simon, 1975, Theorem IV.21, p. 115) implies that the unit ball of $\mathfrak{A}^*$ is compact in the weak* topology, so $\mathcal{P}(\mathfrak{A})$ is a closed subset of a compact set, and hence is also compact in the weak* topology.

Now, define for each $A \in \mathfrak{A}$ the function $\hat{A} \colon \mathcal{P}(\mathfrak{A}) \to \mathbb{C}$ by

$$\hat{A}(\omega) := \omega(A) \tag{2.23}$$

for each $\omega \in \mathcal{P}(\mathfrak{A})$. Notice that $\hat{A}$ is continuous. Furthermore, the algebraic operations in $C(\mathcal{P}(\mathfrak{A}))$ correspond to the algebraic operations in $\mathfrak{A}$: for any $A, B \in \mathfrak{A}$, $\alpha \in \mathbb{C}$, and $\omega \in \mathcal{P}(\mathfrak{A})$, we have

$$\widehat{(A + B)}(\omega) = \omega(A + B) = \omega(A) + \omega(B) = \hat{A}(\omega) + \hat{B}(\omega) = (\hat{A} + \hat{B})(\omega)$$

$$\widehat{(\alpha A)}(\omega) = \omega(\alpha A) = \alpha \cdot \omega(A) = \alpha \cdot \hat{A}(\omega) = (\alpha \hat{A})(\omega)$$

$$\widehat{(AB)}(\omega) = \omega(AB) = \omega(A)\omega(B) = \hat{A}(\omega)\hat{B}(\omega) = (\hat{A}\hat{B})(\omega). \tag{2.24}$$

One can show that the positivity of a linear functional ω implies that $\omega(A^*) = \overline{\omega(A)}$ for all $A \in \mathfrak{A}$, so we similarly have

$$\widehat{(A^*)}(\omega) = \omega(A^*) = \overline{\omega(A)} = \overline{\hat{A}(\omega)} = (\hat{A})^*(\omega). \tag{2.25}$$

Consider the map $\varphi \colon A \mapsto \hat{A}$. We know φ is injective because distinct elements of $\mathfrak{A}$ are always assigned distinct expectation values by some pure state. Further, the Stone–Weierstrass theorem (Reed and Simon, 1975, Theorem IV.10, p. 102) implies that φ is surjective. Thus, $\mathfrak{A}$ is *-isomorphic to $C(\mathcal{P}(\mathfrak{A}))$. $\qquad\square$

This shows that every commutative C*-algebra can be understood as a C*-algebra of functions on a topological space.

We now state the fundamental theorem for noncommutative C*-algebras and sketch a proof.[37]

Theorem 4 (General Gelfand–Naimark) *If $\mathfrak{A}$ is a C*-algebra, then $\mathfrak{A}$ is *-isomorphic to a C*-subalgebra of $\mathcal{B}(\mathcal{H})$ for some Hilbert space $\mathcal{H}$.*

The basic strategy of the proof is to construct a Hilbert space $\mathcal{H}_\omega$ carrying a representation of $\mathfrak{A}$ for each state $\omega \in S(\mathfrak{A})$, and then take $\mathcal{H}$ to be the direct sum of all of the spaces $\mathcal{H}_\omega$. The definition of the Hilbert space $\mathcal{H}_\omega$ follows the

---

[37] See Theorem 4.5.6 and Remark 4.5.8 of Kadison and Ringrose (1997, p. 281).

so-called GNS construction (for Gelfand, Naimark, and Segal). The central idea is to start with the vector space $\mathfrak{A}$ itself – where elements of $\mathfrak{A}$ can be understood as bounded linear operators acting on this vector space by left multiplication – and define a bilinear map from the state ω that can be used to construct an inner product. We state and sketch a proof of the GNS theorem before proceeding.[38]

Theorem 5 (Gelfand–Naimark–Segal) *If $\mathfrak{A}$ is a unital C*-algebra and ω is a state on $\mathfrak{A}$, then there is a Hilbert space $\mathcal{H}_\omega$ with a *-homomorphism $\pi_\omega : \mathfrak{A} \to \mathcal{B}(\mathcal{H}_\omega)$ and a vector $\Omega_\omega \in \mathcal{H}_\omega$ such that for each $A \in \mathfrak{A}$,*

$$\omega(A) = \langle \Omega_\omega, \pi_\omega(A)\Omega_\omega \rangle. \tag{2.26}$$

Proof sketch of Theorem 5: Define a bilinear map $\langle \cdot, \cdot \rangle_0$ on $\mathfrak{A}$ by

$$\langle A, B \rangle_0 := \omega(A^*B) \tag{2.27}$$

for every $A, B \in \mathfrak{A}$. This map is not yet an inner product because it is degenerate – in the sense that it assigns nonzero elements of $\mathfrak{A}$ norm zero. To construct an inner product space, we will quotient out the norm zero vectors.

To that end, let $L_\omega = \{A \in \mathfrak{A} \mid \langle A, A \rangle_0 = 0\}$. Then one can check that L_ω is a closed left ideal in $\mathfrak{A}$, and so the quotient $\mathfrak{A}/L_\omega = \{A + L_\omega \mid A \in \mathfrak{A}\}$ is a vector space with the inner product

$$\langle A + L_\omega, B + L_\omega \rangle := \langle A, B \rangle_0. \tag{2.28}$$

However, $\mathfrak{A}/L_\omega$ may not yet be a Hilbert space because it may fail to be complete relative to this inner product – in the sense that there may be Cauchy sequences of vectors that fail to converge. So, we define $\mathcal{H}_\omega$ as the completion of $\mathfrak{A}/L_\omega$, which can be constructed as a collection of equivalence classes of Cauchy sequences in $\mathfrak{A}/L_\omega$, hence containing $\mathfrak{A}/L_\omega$ as a dense subset (represented as the sequences whose elements are all identical).

With this Hilbert space in hand, we define a representation of $\mathfrak{A}$ on $\mathcal{B}(\mathcal{H}_\omega)$ as follows. For any $A \in \mathfrak{A}$, define the action of $\pi_\omega(A)$ on the dense subset $\mathfrak{A}/L_\omega$ by

$$\pi_\omega(A)(B + L_\omega) := AB + L_\omega. \tag{2.29}$$

One can show that $\pi_\omega(A)$ is continuous, and so this action extends to all of $\mathcal{H}_\omega$.

---

[38] See chapter 4.5 of Kadison and Ringrose (1997) or chapter C.12 of Landsman (2017).

Finally, we must show that π_ω is a *-homomorphism by showing that it preserves the algebraic structure of $\mathfrak{A}$. To see this, note that for every $A, B, C \in \mathfrak{A}$ and $\alpha \in \mathbb{C}$,

$$\pi_\omega(A + B)(C + L_\omega) = (A + B)C + L_\omega$$
$$= (AC + L_\omega) + (BC + L_\omega)$$
$$= (\pi_\omega(A) + \pi_\omega(B))(C + L_\omega)$$
$$\pi_\omega(\alpha A)(C + L_\omega) = (\alpha A)C + L_\omega$$
$$= \alpha(AC + L_\omega)$$
$$= \alpha \pi_\omega(A)(C + L_\omega)$$
$$\pi_\omega(AB)(C + L_\omega) = (AB)C + L_\omega$$
$$= \pi_\omega(A)(BC + L_\omega) \qquad (2.30)$$
$$= \pi_\omega(A)\pi_\omega(B)(C + L_\omega)$$
$$\langle \pi_\omega(A)(B + L_\omega), C + L_\omega \rangle = \langle AB + L_\omega, C + L_\omega \rangle$$
$$= \omega((AB)^* C)$$
$$= \omega(B^* A^* C)$$
$$= \langle B + L_\omega, A^* C + L_\omega \rangle$$
$$= \langle B + L_\omega, \pi_\omega(A^*)(C + L_\omega) \rangle.$$

Since $\mathfrak{A}/L_\omega$ is dense in $\mathcal{H}_\omega$, this implies that for every $A, B \in \mathfrak{A}$

$$\pi_\omega(A + B) = \pi_\omega(A) + \pi_\omega(B)$$
$$\pi_\omega(\alpha A) = \alpha \pi_\omega(A)$$
$$\pi_\omega(AB) = \pi_\omega(A)\pi_\omega(B) \qquad (2.31)$$
$$\pi_\omega(A^*) = \pi_\omega(A)^\dagger.$$

Hence, π_ω is a *-homomorphism.

Lastly, define the vector Ω_ω as the element $I + L_\omega$ in the dense subset $\mathfrak{A}/L_\omega$ of $\mathcal{H}_\omega$. Then for any $A \in \mathfrak{A}$, we have

$$\langle \Omega_\omega, \pi_\omega(A)\Omega_\omega \rangle = \langle I + L_\omega, \pi_\omega(A)(I + L_\omega) \rangle$$
$$= \langle I + L_\omega, AI + L_\omega \rangle$$
$$= \omega(I^* AI) \qquad (2.32)$$
$$= \omega(A).$$

as in the statement of the theorem. $\qquad \square$

This establishes that every state on $\mathfrak{A}$ corresponds to a representation of $\mathfrak{A}$ as operators on a Hilbert space. Finally, we provide an argument sketch that one can always find a representation of $\mathfrak{A}$ that is a *-*isomorphism* to a subset of the bounded operators on a Hilbert space, that is, one can always construct a representation that is *faithful*.

Proof sketch of Theorem 4: Define the *universal representation* π_u of $\mathfrak{A}$ on $\mathcal{H}_u$ by

$$\mathcal{H}_u := \bigoplus_{\omega \in S(\mathfrak{A})} \mathcal{H}_\omega$$

$$\pi_u(A) := \bigoplus_{\omega \in S(\mathfrak{A})} \pi_\omega(A), \tag{2.33}$$

where π_ω is the representation on $\mathcal{H}_\omega$ from the GNS construction in Theorem 5. That π_u is a *-homomorphism follows immediately from the definition of the direct sum.

To see that π_u is a *-isomorphism, one needs to show that for any $A \in \mathfrak{A}$, if $A \neq 0$, then $\pi_u(A) \neq 0$. But one can show that for any nonzero $A \in \mathfrak{A}$, there is a state $\omega_A \in S(\mathfrak{A})$ such that $\omega_A(A) \neq 0$ (Kadison and Ringrose, 1997, Theorem 4.3.4, p. 258). Hence, Theorem 5 implies there is a vector $\Omega_{\omega_A} \in \mathcal{H}_{\omega_A} \subseteq \mathcal{H}_u$ such that

$$\langle \Omega_{\omega_A}, \pi_u(A)\Omega_{\omega_A} \rangle = \langle \Omega_{\omega_A}, \pi_{\omega_A}\Omega_{\omega_A} \rangle = \omega_A(A) \neq 0. \tag{2.34}$$

Therefore, $\pi_u(A) \neq 0$. It follows that π_u is faithful, and hence a *-isomorphism. $\square$

This establishes that every C*-algebra can be understood as an algebra of bounded operators on a Hilbert space.

2.4 Compendium: Examples of C*-Algebras

We now close this brief introduction to the general theory of C*-algebras with a discussion of some of the options for C*-algebras to represent physical quantities in classical and quantum theories. The following is a noncomprehensive list of some of the further C*-algebras that will appear in our discussion of the classical–quantum correspondence.

Classical Algebras

Example 6 (The continuous bounded functions $C_b(X,\tau)$) *Consider a locally compact Hausdorff topological space X with topology τ. Usually X will be the phase space for some physical system endowed with some relevant topology τ. $C_b(X,\tau)$ is the C*-algebra of all bounded, continuous functions $f: X \to \mathbb{C}$.*

$C_b(X, \tau)$ *is more permissive than* $C_0(X, \tau)$ *in the sense that it contains more functions by dropping the requirement that they vanish at infinity. One can see this by examining the state space of* $C_b(X, \tau)$, *which is much bigger than that of* $C_0(X, \tau)$. *Whereas the pure state space of* $C_0(X, \tau)$ *is homeomorphic to* X *as in Theorem 3, the pure state space of* $C_b(X, \tau)$ *is homeomorphic to the Stone–Čech compactification of* X.[39]

Example 7 (The uniformly continuous, bounded functions $UC_b(X, d)$) *Consider a metric space* X *with metric* d. *Again, usually* X *will be the phase space for some physical system endowed with some relevant metric that generates a topology.* $UC_b(X, d)$ *is the C*-algebra of all bounded, uniformly continuous functions* $f: X \to \mathbb{C}$. $UC_b(X, d)$ *is, in general, smaller than* $C_b(X, \tau_d)$, *but larger than* $C_0(X, \tau_d)$, *and so it serves as an intermediate algebra. We do not know of a nice characterization of the pure state space of* $UC_b(X, d)$, *but we expect it to be related to the completion of* X *relative to the metric* d.[40]

Example 8 (The bounded measurable functions $B(X, \Sigma)$) *Consider a measurable space* X *with* σ*-algebra* Σ. *Again, usually* X *will be the phase space for some physical system endowed with some relevant topology, generating the* σ*-algebra* Σ. $B(X, \Sigma)$ *is the collection of all bounded, measurable functions* $f: X \to \mathbb{C}$.[41] $B(X, \Sigma)$ *is a C*-algebra that is even more permissive than the previous examples. It contains many functions, including discontinuous functions and even characteristic functions, which are projections (idempotent elements) in the algebra. We again do not know of a nice characterization of the pure state space of* $B(X, \Sigma)$, *but it will again contain all points in* X *as well as additional states.*

Example 9 (The almost periodic functions $AP(V')$) *Consider a topological vector space* V *and the dual space* V' *of all continuous linear functionals on* V. *In general, we understand* V' *as the phase space of a system and* V *as the space of "test functions." If* V' *is the phase space of an unconstrained system with finitely many degrees of freedom so that* $V' = \mathbb{R}^{2n}$, *then we also have* $V = \mathbb{R}^{2n}$. *But* V *might also be the space of possible test functions for a linear field theory, for example, for a scalar field theory, we might choose the space* $V = \mathcal{S}(\mathbb{R}^d)$ *of*

---

[39] See also Gamelin (1969).

[40] See also Aliprantis and Border (1999).

[41] Recall that a function $f: X \to \mathbb{C}$ is *measurable* with respect to a σ-algebra Σ for X if for every measurable set $O \subseteq \mathbb{C}$, the set $f^{-1}[O]$ is measurable in X.

Schwartz functions on a flat d-dimensional space,[42] *so that* $V' = S'(\mathbb{R}^d)$ *is the phase space for the theory containing both smooth and singular field configurations. Consider, for each test function* $f \in V$, *the function* $W_0(f): V' \to \mathbb{C}$ *defined by*

$$W_0(f)(F) := e^{iF[f]} \tag{2.35}$$

for each functional $F \in V'$. $AP(V')$ *is the C*-algebra of continuous almost periodic functions, which can be understood as generated from the functions* $W_0(f)$ *as a basis and completed in the supremum norm.* $AP(V')$ *is a subset of* $C_b(V', \tau)$, *and the pure state space of* $AP(V')$ *is homeomorphic to what is known as the Bohr compactification of* V'.[43]

Example 10 (The commutative resolvent algebra $C_{\mathcal{R}}(V')$) *Consider a topological vector space* V *and the dual space* V' *of all continuous linear functionals on* V. *Again,* V *will usually represent the space of test functions for a theory with phase space* V', *for example,* $V = V' = \mathbb{R}^{2n}$ *for systems with finitely many degrees of freedom. Consider, for each* $\lambda \in \mathbb{R} \setminus \{0\}$ *and* $f \in V$, *the function* $R_0(\lambda, f): V' \to \mathbb{C}$ *defined by*

$$R_0(\lambda, f)(F) := \frac{1}{i\lambda - F[f]} \tag{2.36}$$

for all $F \in V'$. $C_{\mathcal{R}}(V')$ *is the C*-algebra of classical resolvents, which can be understood as generated freely from the functions* $R_0(\lambda, f)$, *ranging over each* $\lambda \in \mathbb{R} \setminus \{0\}$ *and* $f \in V$. *This classical resolvent algebra is contained in* $C_b(V', \tau)$, *and its pure state space corresponds to the Grassmanian of* V' *understood as a compactification of* V'.[44]

Quantum Algebras

Example 11 (The bounded operators $\mathcal{B}(\mathcal{H})$) *Consider a Hilbert space* $\mathcal{H}$. *Usually,* $\mathcal{H}$ *will be the Hilbert space associated to some quantum system.* $\mathcal{B}(\mathcal{H})$ *is the C*-algebra of all bounded linear operators on* $\mathcal{H}$. $\mathcal{B}(\mathcal{H})$ *is more permissive than* $\mathcal{K}(\mathcal{H})$ *in that it contains many more operators by dropping the requirement of compactness. For example,* $\mathcal{B}(\mathcal{H})$ *contains all projection operators (whereas* $\mathcal{K}(\mathcal{H})$ *contains finite-dimensional projections), and so it*

---

[42] Usually we choose $d = 3$ or 4, depending on whether one works with the full configuration space or a space of initial conditions on a spacelike surface.

[43] See also Anzai and Kakutani (1943), Hewitt (1953), or Binz et al. (2004a).

[44] See also van Nuland (2019).

provides a rich enough setting to formulate spectral theory. Moreover, while every density operator on $\mathcal{H}$ defines a state on $\mathcal{B}(\mathcal{H})$, the algebra also contains many singular states that cannot be represented as density operators on $\mathcal{H}$.

Example 12 (The Weyl algebra $\mathcal{W}(V, \hbar\sigma)$) *Consider a symplectic topological vector space V with symplectic form σ.[45] As previously, V will usually represent the space of test functions for a theory with phase space V'. Consider the C^*-algebra generated by the basis elements $W_\hbar(f)$ for $f \in V$ with operations*

$$W_\hbar(f)W_\hbar(g) := e^{\frac{-i\hbar\sigma(f,g)}{2}} W_\hbar(f+g)$$
$$W_\hbar(f)^* := W_\hbar(-f) \tag{2.37}$$

for $f, g \in V$ and some fixed real number $\hbar > 0$. There is a unique norm satisfying the C^*-identity and compatible with these algebraic operations. The Weyl algebra $\mathcal{W}(V, \hbar\sigma)$ is the C^*-algebra generated by the elements $W_\hbar(f)$. In the special case where $V = \mathbb{R}^{2n}$, one can understand $\mathcal{W}(V, \hbar\sigma)$ through the standard Schrödinger representation π_S on $\mathcal{H}_S = L^2(\mathbb{R}^n)$. In this case, if we let Q_j and P_j denote the position and momentum operators

$$(Q_j\psi)(x) := x_j\psi(x)$$
$$(P_j\psi)(x) := i\frac{\partial}{\partial x_j}\psi(x) \tag{2.38}$$

for all $\psi \in L^2(\mathbb{R})$, then π_S is the continuous linear extension of the representation[46]

$$\pi_S(W_\hbar(a,b)) := e^{i\sum_{j=1}^{n} a_j\sqrt{\hbar}\cdot P_j + ib_j\sqrt{\hbar}\cdot Q_j} \tag{2.39}$$

so that $\mathcal{W}(V, \hbar\sigma)$ can be understood as the C^*-algebra generated by exponentials of configuration and momentum quantities.[47]

Example 13 (The resolvent algebra $\mathcal{R}(V, \hbar\sigma)$) *Consider a symplectic topological vector space V with symplectic form σ. Again, V will usually represent*

---

[45] Recall that a map $\sigma: V \times V \to \mathbb{R}$ is a *symplectic form* if it is bilinear, antisymmetric in the sense that $\sigma(u, v) = -\sigma(v, u)$ for all $u, v \in V$, and nondegenerate in the sense that if $\sigma(u, v) = 0$ for all $v \in V$, then $u = 0$.

[46] Note that we do not include any factor of $\hbar$ in our definition of Q_j and P_j in Eq. (2.38). In this sense, our definition of the momentum operator does not agree with what is sometimes called the standard momentum operator. Instead, we opt to include factors of $\hbar$ in our definition of π_S, and here we make a particular choice, which follows the convention of van Nuland (2019) rather than that of Landsman (1998a) to maintain symmetry in interchanging Q_j and P_j. Our motivation for making this choice is ultimately to set up for the discussion of the effect of unit changes on elements of the Weyl algebra in Section 4.

[47] See also Slawny (1972), Manuceau et al. (1974), Petz (1990), or Binz et al. (2004a). For an introduction for philosophers, see Clifton and Halvorson (2001).

the space of test functions for a theory with phase space V'. Consider the C*-algebra generated by the elements $R_\hbar(\lambda, f)$ for $\lambda \in \mathbb{R} \setminus \{0\}$ and $f \in V$ with operations

$$R_\hbar(\lambda, 0) := -\frac{i}{\lambda} I$$

$$R_\hbar(\lambda, f)^* := R_\hbar(-\lambda, f)$$

$$\nu R_\hbar(\nu\lambda, \nu f) := R_\hbar(\lambda f)$$

$$R_\hbar(\lambda, f) - R_\hbar(\mu, f) := i(\mu - \lambda) R_\hbar(\lambda, f) R_\hbar(\mu, f) \tag{2.40}$$

$$[R_\hbar(\lambda, f), R_\hbar(\mu, g)] := i\hbar\sigma(f, g) R_\hbar(\lambda, f) R_\hbar(\mu, g)^2 R_\hbar(\lambda, f)$$

$$R_\hbar(\lambda, f) R_\hbar(\mu, g) := R_\hbar(\lambda + \mu, f + g) [R_\hbar(\lambda, f) + R_\hbar(\mu, g)$$
$$+ i\hbar\sigma(f, g) R_\hbar(\lambda, f)^2 R_\hbar(\mu, g)]$$

for $f, g \in V$, $\lambda, \mu, \nu \in \mathbb{R} \setminus \{0\}$ with $\lambda + \mu \neq 0$, and some fixed real number $\hbar > 0$. As with the Weyl algebra, in the special case where $V = \mathbb{R}^{2n}$, one can understand $\mathcal{R}(V, \hbar\sigma)$ through the standard Schrödinger representation, which we will again denote π_S on $\mathcal{H}_S = L^2(\mathbb{R}^n)$. In this case, π_S is the continuous linear extension of the representation

$$\pi_S(R_\hbar(\lambda, (a, b))) := \left(i\lambda - \left(\sum_{j=1}^n a_j \sqrt{\hbar} \cdot P_j + b_j \sqrt{\hbar} \cdot Q_j \right) \right)^{-1} \tag{2.41}$$

so that $\mathcal{R}(V, \hbar\sigma)$ can be understood as the C*-algebra generated by resolvents of configuration and momentum quantities.[48]

Example 14 (The (reduced) group C*-algebra $C^*(G)$) *Consider a finite-dimensional Lie group G.*[49] *We will assume G is unimodular, which means that it possesses an invariant measure dx called the Haar measure. Consider the Hilbert space $L^2(G)$ of functions $\psi: G \to \mathbb{C}$ with $\int_G |\psi(x)|^2 \, dx < \infty$. Notice that a function $f \in C_c^\infty(G)$ that is smooth and compactly supported corresponds to an operator A_f on $L^2(G)$ acting by convolution as*

$$(A_f \psi)(x) = \int_G f(y) \psi(y^{-1} x) \, dy \tag{2.42}$$

for all $\psi \in L^2(G)$ and $x \in G$. The (reduced) group C*-algebra $C^*(G)$ is the C*-algebra acting on $L^2(G)$ generated by the operators A_f. This C*-algebra

---

[48] See also Buchholz and Grundling (2008, 2015), Buchholz (2017, 2018), and van Nuland (2019).

[49] A Lie group G is a group that is also a manifold, whose group operation is a smooth map from $G \times G \to G$, and whose inverse operation is a smooth map $G \to G$.

can be understood as the completion in a suitable norm of $C_c^\infty(G)$[50] with the algebraic operations

$$(fg)(x) = \int_G f(y)g(y^{-1}x)\, dy$$

$$f^*(x) = \overline{f(x^{-1})}$$

(2.43)

for $f, g \in C_c^\infty(G)$ and $x \in G$. The algebra $C^*(G)$ can be used to represent bounded functions of (generally unbounded) charge quantities associated with a symmetry group G. We mention without proof that representations of $C^*(G)$ as bounded operators on a Hilbert space correspond to unitary representations of G itself.[51]

Example 15 (The (reduced) transformation group C*-algebra $C^*(G, Q)$) *Consider a locally compact Lie group G acting (on the left) on a manifold Q. We again assume G is unimodular with Haar measure dx and we similarly suppose that Q carries a measure dq whose support is the whole space Q. Consider the Hilbert space $L^2(G \times Q)$ of functions $\psi : G \times Q \to \mathbb{C}$ with $\int_G \int_Q |\psi(x, q)|^2\, dx dq < \infty$. Notice that a function $f \in C_c^\infty(G)$ corresponds to an operator A_f on $L^2(G \times Q)$ acting by convolution as*

$$(A_f\psi)(x, q) = \int_G f(y, xq)\psi(y^{-1}x, q)\, dy$$

(2.44)

for $\psi \in L^2(G \times Q)$, $x \in G$, and $q \in Q$. The transformation group C*-algebra $C^*(G, Q)$ is the C*-algebra acting on $L^2(G \times Q)$ generated by the operators A_f. This C*-algebra can be understood as the completion in a suitable norm of $C_c^\infty(G \times Q)$[52] with the algebraic operations

$$(fg)(x, q) = \int_G f(y, q)g(y^{-1}x, y^{-1}q)\, dy$$

---

[50] Both the reduced group C*-algebra and the full group C*-algebra are generated by the same set of elements considered here, but in general they constitute completions with respect to different C*-norms. When G is *amenable*, that is, it possesses an invariant mean, the reduced group C*-algebra is *-isomorphic to the group C*-algebra. For quantization, one need not restrict attention to amenable groups; however, we will ignore the technical subtleties here and refer the reader especially to Landsman (2017, 1999) for full generality. In the main text, we restrict attention for simplicity only to amenable groups so that we do not need to distinguish these algebras.

[51] See chapter C.18 of Landsman (2017). This last claim applies to the full group C*-algebra.

[52] As in fn. 50, one can define both a reduced transformation group C*-algebra and a full transformation group C*-algebra, which are generated by the same set of elements considered here, but in general they constitute completions with respect to different C*-norms. Again, if G is *amenable*, then these algebras are *-isomorphic. As before, for quantization, one need not restrict attention to amenable groups; however, we will ignore the technical subtleties here and refer the reader especially to Landsman (2017, 1999) for full generality. In the main text, we restrict attention for simplicity only to amenable groups so that we do not need to distinguish these algebras.

$$f^*(x,q) = \overline{f(x^{-1}, x^{-1}q)} \qquad\qquad (2.45)$$

for f,g ∈ $C_c^\infty(G \times Q)$, $x \in G$, and $q \in Q$. The algebra $C^*(G,Q)$ can be used to represent the quantities of a quantum system moving in the configuration space Q with charge associated with the symmetry group G. Clearly $C^*(G)$ is a special case of $C^*(G,Q)$ when Q is trivial. We also mention without proof that the algebra $\mathcal{K}(L^2(G))$ is *-isomorphic to the special case $C^*(G,G)$ when we consider the action of G on itself by left multiplication.[53]

3 Quantization and the Classical Limit

Now that we have the C*-algebraic framework in hand to provide common formulations for classical and quantum physics, we will proceed to develop mathematical tools for comparing the two. Our starting point will be the notion of a *quantization map*, which provides a way of identifying a mathematical object as representing the same physical quantity in different algebras. We will use these quantization maps to develop the theory of strict quantization, whose basic mathematical structure is known as a *continuous bundle of C*-algebras*. This setting will allow us to investigate both the construction of quantum theories from classical starting points, as well as the dual process of taking the classical $\hbar \to 0$ limit of a quantum theory.[54] At the end of the section, we also discuss the relation between quantization maps that we use and other approaches to the construction of quantum theories.[55]

---

[53] See chapter C.18 of Landsman (2017). This last claim applies to the full transformation group C*-algebra.

[54] Of course, the $\hbar \to 0$ limit is only one example of a classical limit. While $\hbar \to 0$ is the only classical limit we will treat here, one can use similar mathematical tools to treat other macroscopic limits like the $N \to \infty$ limit for composite systems (see Landsman, 2007, 2017).

[55] We have made a significant choice in the approaches to quantization we consider here. We will omit discussion of four other approaches to the quantum–classical relation:

- First, there is a prominent notion of quantization called *geometric quantization*, which uses rather different mathematical tools. For an introduction to geometric quantization, see Woodhouse (1997) or the discussion and references in Landsman (2007). See also Emch (1983) for more on the relation between geometric quantization and the classical limit.

- Second, there is the field of *semiclassical analysis*, which approaches many of the same issues we deal with here in the $\hbar \to 0$ limit, but from the mathematical perspective of differential equations and microlocal analysis. See Martinez (2002) or Zworski (2012) for introductions.

- Third, the *quantization–dequantization* approach of Gracia-Bondía (1992) is closely related to our discussion of Weyl and Berezin quantization that follows, but emphasizes different mathematical aspects.

- Fourth, there is *path integral quantization*, which uses somewhat different tools developed originally by Feynmann and motivated especially by applications to quantum field theory. These tools are also related to some of the discussion of asymptotics and semiclassical trace formulae mentioned later in §4.2. See, for example, Mazzucchi (2009).

3.1 Strict Deformation Quantization

To begin, suppose one has a classical theory formulated on a phase space given by a Poisson manifold $\mathcal{M}$. Further, suppose that $\mathcal{P}$ is a Poisson algebra of functions on $\mathcal{M}$, that is, an associative algebra equipped with the Poisson bracket acting as a Lie bracket.

Definition 5 A *strict quantization* of $\mathcal{P}$ consists in a family of C*-algebras $(\mathfrak{A}_\hbar)_{\hbar \in [0,1]}$ and a family of linear, *-preserving *quantization maps* $(Q_\hbar : \mathcal{P} \to \mathfrak{A}_\hbar)_{\hbar \in [0,1]}$, where for each $\hbar \in [0, 1]$, the image $Q_\hbar[\mathcal{P}]$ is dense in $\mathfrak{A}_\hbar$ and Q_0 is the identity. The maps $Q_\hbar$ are required to satisfy the following constraints for all $A, B \in \mathcal{P}$:

(i) (*Dirac's condition*): $\lim_{\hbar \to 0} \|\frac{i}{\hbar}[Q_\hbar(A), Q_\hbar(B)] - Q_\hbar(\{A, B\})\|_\hbar = 0$;
(ii) (*von Neumann's condition*): $\lim_{\hbar \to 0} \|Q_\hbar(A)Q_\hbar(B) - Q_\hbar(AB)\|_\hbar = 0$; and
(iii) (*Rieffel's condition*): the function $\hbar \mapsto \|Q_\hbar(A)\|_\hbar$ is continuous.

Here, $\|\cdot\|_\hbar$ denotes the norm in the C*-algebra $\mathfrak{A}_\hbar$. If, furthermore, for each $\hbar \in [0, 1]$, $Q_\hbar$ is faithful and $Q_\hbar[\mathcal{P}]$ is closed under algebraic operations, then $(\mathfrak{A}_\hbar, Q_\hbar)_{\hbar \in [0,1]}$ is called a *strict deformation quantization*.[56]

The conditions of a strict quantization capture the idea that in the limit as $\hbar \to 0$, the algebraic operations of the C*-product and commutator Lie bracket from the quantum theory approach the pointwise product and Poisson bracket of the classical theory.[57] The deformation condition further allows one to pull back the product for each $\hbar > 0$ to $\mathcal{P}$ and define a family of products $\star_\hbar : \mathcal{P} \times \mathcal{P} \to \mathcal{P}$ by

$$A \star_\hbar B := Q_\hbar^{-1}(Q_\hbar(A)Q_\hbar(B)), \tag{3.1}$$

thus representing the algebraic structure of a quantum theory via phase space functions.

Given a strict deformation quantization, one has the tools to discuss classical limits of both states and quantities. The quantity $A \in \mathcal{P}$ is understood as the classical limit of the family of quantities $Q_\hbar(A)$ for $\hbar > 0$. Moreover, one defines a *continuous field of states* to be a family $(\omega_\hbar)_{\hbar \in [0,1]}$ of states, each on the corresponding C*-algebra $\mathfrak{A}_\hbar$ satisfying the condition that for each $A \in \mathcal{P}$,

---

[56] See Ch. II of Landsman (1998a), Landsman (2007), and Rieffel (1989, 1993).

[57] A number of "no-go" theorems preclude the identification of the Poisson bracket with $\frac{i}{\hbar}$ times the commutator exactly for all classical observables (Groenewold, 1946; van Hove, 1951). This motivates the statement of Dirac's condition in terms of this identity holding only approximately in the $\hbar \to 0$ limit. See also Gotay (1980, 1999) for discussion of the "no-go" theorems and other approaches.

the map $\hbar \mapsto \omega_\hbar(Q_\hbar(A))$ is continuous. Given a continuous field of states $(\omega_\hbar)_{\hbar \in [0,1]}$, the classical state ω_0 is understood as the classical limit of the family of states $\omega_\hbar$ for $\hbar > 0$.

In general, there may be many different families of quantization maps for a given Poisson algebra $\mathcal{P}$. Two families of quantization maps $(Q_\hbar)_{\hbar \in [0,1]}$ and $(Q'_\hbar)_{\hbar \in [0,1]}$ for $\mathcal{P}$ with the same family of C*-algebras $(\mathfrak{A}_\hbar)_{\hbar \in [0,1]}$ are called *equivalent* if, for each $A \in \mathcal{P}$, the map

$$\hbar \mapsto \|Q_\hbar(A) - Q'_\hbar(A)\|_\hbar \tag{3.2}$$

is continuous and hence approaches zero in the limit as $\hbar \to 0$ (since $Q_0(A) = Q'_0(A)$). In this case, the maps $Q_\hbar$ and $Q'_\hbar$ capture the same information in the limit as $\hbar \to 0$ for the algebraic structure of the physical quantities.

In fact, one can find a common mathematical structure that is shared among equivalent quantization maps. This structure, called a *continuous bundle of C*-algebras*, provides just what is needed to determine the classical limit of a quantum theory (as we will see later).

Definition 6 A *continuous bundle of C*-algebras* over a locally compact topological space I consists in:

- a C*-algebra $\mathfrak{A}$ called the algebra of *continuous sections*;
- a family of C*-algebras $(\mathfrak{A}_\hbar)_{\hbar \in I}$ called the *fibers*; and
- a family of surjective *-homomorphisms $\phi_\hbar: \mathfrak{A} \to \mathfrak{A}_\hbar$ called the *evaluation maps*.

Together, these structures must satisfy:

(i) for each $a \in \mathfrak{A}$, $\|a\| = \sup_{\hbar \in I} \|\phi_\hbar(a)\|_\hbar$;
(ii) for each $f \in C_0(I)$ and each $a \in \mathfrak{A}$, there is a section $fa \in \mathfrak{A}$ such that $\phi_\hbar(fa) = f(\hbar)\phi_\hbar(a)$ for each $\hbar \in I$;
(iii) for each section $a \in \mathfrak{A}$, the map $\hbar \mapsto \|a\|_\hbar$ belongs to $C_0(I)$.[58]

Although we allow for a wider array of base spaces, most often we take I to be the interval $[0, 1]$, representing the possible values of $\hbar$. The fibers are, as before, the C*-algebras of physical quantities for either classical or quantum theories at different values of $\hbar$. The continuous sections can be understood as maps that assign to each point $\hbar$ in the base space I an element of the fiber $\mathfrak{A}_\hbar$ above that point. The evaluation map $\phi_\hbar$ then yields, for each section a, the

---

[58] See also chapter 10 of Dixmier (1977), chapter II of Landsman (1998a), chapter C.19 of Landsman (2017), and Kirchberg and Wasserman (1995).

value $\phi_\hbar(a)$ at the point $\hbar$ in the base space. When we understand sections in this way, we will sometimes denote them by $[\hbar \mapsto \phi_\hbar(a)]$ to emphasize their character as maps from the base space to the fibers.

Each strict quantization generates a continuous bundle of C*-algebras, the continuous sections of which contain the trajectories of the quantization maps. This is captured by the following proposition.

Proposition 1 *Suppose $\mathcal{P}$ is a Poisson algebra. Suppose $(\mathfrak{A}_\hbar, Q_\hbar)_{\hbar \in [0,1]}$ is a strict quantization such that the map $\hbar \mapsto \|P(\hbar)\|_\hbar$ is continuous for each polynomial[59] P of elements of the form $[\hbar \mapsto Q_\hbar(A)]$ for $A \in \mathcal{P}$. Then there is a unique (up to *-isomorphism) continuous bundle of C*-algebras $((\mathfrak{A}_\hbar, \phi_\hbar)_{\hbar \in [0,1]}, \mathfrak{A})$ such that for each $A \in \mathcal{P}$, there is a section $a \in \mathfrak{A}$ with*

$$\phi_\hbar(a) = Q_\hbar(A). \tag{3.4}$$

for each $\hbar \in [0,1]$. We call $\mathfrak{A}$ the algebra of continuous sections generated by the quantization maps $Q_\hbar$.[60]

The proof of Proposition 1 constructs $\mathfrak{A}$ as follows. We define $\mathfrak{A}$ as the set of all $a \in \prod_{\hbar \in I} \mathfrak{A}_\hbar$ such that for each $A \in \mathcal{P}$, the map

$$\hbar \mapsto \|a(\hbar) - Q_\hbar(A)\|_\hbar \tag{3.5}$$

is in $C_0(I)$. So defined, $\mathfrak{A}$ is a C*-algebra, and with the evaluation maps

$$\phi_\hbar(a) := a(\hbar) \tag{3.6}$$

it becomes the algebra of continuous sections for a continuous bundle of C*-algebras.

It follows immediately from the uniqueness clause of Proposition 1 that if two quantization maps are equivalent, then they generate the same algebra of continuous sections for a continuous bundle of C*-algebras.

3.2 Compendium: Examples of Strict Quantizations

We now present a number of examples of strict quantizations that will be discussed in the remaining sections. First, we review the two standard quantization procedures on finite-dimensional flat phase spaces for a basic class of functions.

---

[59] More precisely, we mean that $P: [0,1] \to \coprod_{[0,1]} \mathfrak{A}_\hbar$ has the form

$$P(\hbar) = Q_\hbar(A_1) \cdot \ldots \cdot Q_\hbar(A_n) \tag{3.3}$$

for some fixed finite sequence $A_1, \ldots, A_n \in \mathcal{P}$, or that P is a linear combination of sections of the form (3.3).

[60] See Theorem 1.2.4 of Landsman (1998a).

In what follows, we consider $C_c^\infty(\mathbb{R}^{2n})$, the collection of smooth, compactly supported functions on $\mathbb{R}^{2n}$.

Example 16 (Weyl quantization on $C_c^\infty(\mathbb{R}^{2n})$) *Let $\mathcal{M} = \mathbb{R}^{2n}$ with $\mathcal{P} = C_c^\infty(\mathbb{R}^{2n})$, and define the standard Poisson bracket*

$$\{f,g\} := \sum_{j=1}^{n} \frac{\partial f}{\partial q_j} \frac{\partial g}{\partial p_j} - \frac{\partial f}{\partial p_j} \frac{\partial g}{\partial q_j} \tag{3.7}$$

for all $f,g \in C_c^\infty(\mathbb{R}^{2n})$ and canonical coordinates $(q_1,\ldots,q_n,p_1,\ldots,p_n)$ on $\mathbb{R}^{2n}$. Define the fiber algebras by

$$\begin{aligned}
\mathfrak{A}_0 &= C_0(\mathbb{R}^{2n}) \\
\mathfrak{A}_\hbar &= \mathcal{K}(L^2(\mathbb{R}^n)) \quad \text{for } \hbar > 0.
\end{aligned} \tag{3.8}$$

Define the Weyl quantization maps $Q_\hbar^W : C_c^\infty(\mathbb{R}^{2n}) \to \mathcal{K}(L^2(\mathbb{R}^n))$ for $\hbar > 0$ by

$$(Q_\hbar^W(f)\,\psi)(x) := \int_{\mathbb{R}^{2n}} \frac{d^n p\, d^n q}{(2\pi\hbar)^n}\, e^{ip\cdot(x-q)/\hbar}\, f\!\left(p, \frac{1}{2}(x+q)\right) \psi(q) \tag{3.9}$$

for all $\psi \in L^2(\mathbb{R}^n)$ and $f \in C_c^\infty(\mathbb{R}^{2n})$. Then $Q_\hbar^W$ is a strict deformation quantization.[61]

Example 17 (Berezin quantization on $C_c^\infty(\mathbb{R}^{2n})$) *Let $\mathcal{M} = \mathbb{R}^{2n}$ with $\mathcal{P} = C_c^\infty(\mathbb{R}^{2n})$ and the standard Poisson bracket in Eq. (3.7). Define the fiber algebras by*

$$\begin{aligned}
\mathfrak{A}_0 &= C_0(\mathbb{R}^{2n}) \\
\mathfrak{A}_\hbar &= \mathcal{K}(L^2(\mathbb{R}^n)) \quad \text{for } \hbar > 0.
\end{aligned} \tag{3.10}$$

Consider the family of coherent states $\phi_\hbar^{(p,q)} \in L^2(\mathbb{R}^n)$, each centered at a point $(p,q) \in \mathbb{R}^{2n}$, defined by

$$\phi_\hbar^{(p,q)}(x) := (\pi\hbar)^{-n/4} e^{-\frac{ip\cdot q}{2\hbar}} e^{\frac{ip\cdot x}{\hbar}} e^{-\frac{(x-q)^2}{2\hbar}}. \tag{3.11}$$

Define the Berezin quantization maps $Q_\hbar^B : C_c^\infty(\mathbb{R}^{2n}) \to \mathcal{K}(L^2(\mathbb{R}^n))$ for $\hbar > 0$ by

$$(Q_\hbar^B(f)\,\psi)(x) := \int_{\mathbb{R}^{2n}} \frac{d^n p\, d^n q}{(2\pi\hbar)^n}\, \langle \psi, \phi_\hbar^{(p,q)} \rangle\, \phi_\hbar^{(p,q)}(x) \tag{3.12}$$

for all $\psi \in L^2(\mathbb{R}^n)$ and $f \in C_c^\infty(\mathbb{R}^{2n})$. Then $Q_\hbar^B$ is a strict deformation quantization.[62]

---

[61] See chapter II.2.2.5 of Landsman (1998a).
[62] See chapter II.2.2.3 of Landsman (1998a).

One can show that $Q_\hbar^W$ fails to be continuous, and as a result fails to send positive functions to positive operators. On the other hand, $Q_\hbar^B$ does send positive functions to positive operators, and it follows that $Q_\hbar^B$ is continuous. However, the strict deformation quantizations defined by $Q_\hbar^W$ and $Q_\hbar^B$ on $C_c^\infty(\mathbb{R}^{2n})$ are equivalent, and hence determine the same continuous bundle of C*-algebras.

We note in passing that Weyl quantization can be generalized to curved phase spaces that are Riemannian manifolds, while Berezin quantization can be generalized to curved phase spaces that are Kähler manifolds.

For the moment, we restrict our attention to linear phase spaces. Even in this case, one can obtain other strict deformation quantizations by generalizing the Weyl and Berezin quantization procedures to different classes of functions on a phase space. The following examples construct Weyl and Berezin quantization maps on different domains in a way that allows the phase space to be infinite-dimensional as long as it is still linear – hence, allowing for the quantization of linear field theories.

In what follows, we let V be a symplectic topological vector space. For example, for the theory of a scalar field satisfying the Klein–Gordon equation,[63] one can let $V = C_c^\infty(\mathbb{R}^3) \oplus C_c^\infty(\mathbb{R}^3)$ be the space of pairs of test functions (f,g) on space. The dual space V' represents the space of pairs (π, φ) of initial data for field momenta π and field configurations φ, which act on V as linear functionals by smearing fields with test functions as

$$(\pi, \varphi)[f,g] := \int_{\mathbb{R}^3} \pi(x)f(x) + \varphi(x)g(x)\, d^3x. \tag{3.13}$$

Note that this formulation allows one to consider even singular (distributional) field configurations in addition to smooth fields, as required in quantum field theory. The conserved current of the Klein–Gordon equation determines a symplectic form $\sigma: V \times V \to \mathbb{R}$ by

$$\sigma((f,g),(\tilde{f},\tilde{g})) := \int_{\mathbb{R}^3} f(x)\tilde{g}(x) - \tilde{f}(x)g(x)\, d^3x \tag{3.14}$$

for all $(f,g),(\tilde{f},\tilde{g}) \in V$.

As another example, one can take the electromagnetic fields satisfying the vacuum Maxwell equations. Consider the phase space formulation with canonical variables A_a and E^b representing the magnetic 3-vector potential and the electric field 3-vector.[64] In this case, we define V as the collection of pairs of smooth vector fields f^a and covector fields g_b, each on $\mathbb{R}^3$, understood as test

---

[63] See, for example, Halvorson (2001b).

[64] See, for example, Ashtekar and Isham (1992).

functions for the variables A_a and E^b, respectively. We require that for each pair $(f^a, g_b) \in V$, the test functions are compactly supported and divergence free (i.e., $\nabla_a f^a = \nabla^b g_b = 0$). The dual space V' represents the space of pairs (A_a, E^b) of initial data for magnetic vector potentials A_a in the Coulomb gauge (i.e., satisfying the constraint $\nabla^a A_a = 0$) and electric field configurations E^b. We understand the fields to act on the test functions in V as linear functionals by smearing fields with test functions as

$$(A_a, E^b)[f^a, g_b] := \int_{\mathbb{R}^3} A_a(x) f^a(x) + E^b(x) g_b(x) \, d^3 x. \tag{3.15}$$

Note that in the phase space formulation, the electric field configuration E^b is the canonically conjugate momentum to the magnetic vector potential A_a and the constraint that test functions g_b be divergence free implies that one of the source-free Maxwell equations (Gauss's law) $\nabla_a E^a = 0$ is satisfied. The remaining source-free Maxwell equations reduce to the massless Klein–Gordon equation for each component of A_a. This implies that the conserved current defines a symplectic form $\sigma : V \times V \to \mathbb{R}$ by

$$\sigma((f^a, g_b), (\tilde{f}^a, \tilde{g}_b)) := \int_{\mathbb{R}^3} f^a(x) \tilde{g}_a(x) - \tilde{f}^a(x) g_a(x) \, d^3 x \tag{3.16}$$

for all $(f^a, g_b), (\tilde{f}^a, \tilde{g}_b) \in V$.

Thus far, we have established that there are a number of examples of free field theories whose phase space constitutes the dual to a symplectic topological vector space, even though such a space will typically be infinite dimensional. We now consider the quantization of such a theory. We abstract away from the details of any particular field theory and consider just the quantization of functions on the space of initial field configurations.

Example 18 (Weyl quantization on $AP(V')$) *Let V be a symplectic topological vector space with $M = V'$. Define the Poisson algebra $\mathcal{P} = \Delta(V) \subseteq AP(V')$ as the collection of finite linear combinations of the functions $W_0(f)$ on V' (see Eq. (2.35)) ranging over all $f \in V$. Define the Poisson bracket as the bilinear extension of*

$$\{W_0(f), W_0(g)\} := \sigma(f, g) W_0(f + g) \tag{3.17}$$

for all $f, g \in V$. Define the fiber algebras by

$$\begin{aligned} \mathfrak{A}_0 &= AP(V') \\ \mathfrak{A}_\hbar &= \mathcal{W}(V, \hbar \sigma) \quad \text{for } \hbar > 0. \end{aligned} \tag{3.18}$$

Define the Weyl quantization maps $Q_\hbar^W : \Delta(V) \to \mathcal{W}(V, \hbar\sigma)$ *for* $\hbar > 0$ *by the linear extension of*[65]

$$Q_\hbar^W(W_0(f)) := W_\hbar(f) \tag{3.19}$$

for all $f \in V$. *Then* $Q_\hbar^W$ *is a strict deformation quantization.*[66]

In the special case where $V = \mathbb{R}^{2n}$, there is a relationship between Weyl quantization on $C_c^\infty(\mathbb{R}^{2n})$ and $AP(\mathbb{R}^{2n})$ that allows one to think of them as the same quantization procedure. Although the map $Q_\hbar^W$ is not continuous, and so it cannot be continuously extended from $C_c^\infty(\mathbb{R}^{2n})$ to $AP(\mathbb{R}^{2n})$, one can instead treat the operators $W_\hbar(a, b)$ in the Schrödinger representation as exponentials $\pi_S(W_\hbar(a, b)) = e^{i \sum_{j=1}^n a_j \sqrt{\hbar} \cdot P_j + b_j \sqrt{\hbar} \cdot Q_j}$ and think of them as modes in a Fourier decomposition. One can show that for any $f \in C_c^\infty(\mathbb{R}^{2n})$,

$$Q_\hbar^W(f) = \int_{\mathbb{R}^{2n}} \frac{d^n a \, d^n b}{(2\pi\hbar)^n} (\mathcal{F}f)(a, b) \, \pi_S(W_\hbar(a, b)), \tag{3.20}$$

where $\mathcal{F}f$ is the Fourier transform of f and, as above, $Q_\hbar^W(W_0(a, b)) = W_\hbar(a, b)$ is the Weyl quantization map on $AP(\mathbb{R}^{2n})$. This, in some sense, justifies us in referring to both maps with the same symbol.[67]

Since the Berezin quantization map $Q_\hbar^B$ is continuous, one can immediately extend it to larger algebras of functions on the phase space $\mathbb{R}^{2n}$. However, it is useful to provide an explicit form for the Berezin quantization map on $AP(\mathbb{R}^{2n})$ in order to generalize this to infinite-dimensional phase spaces. To do so, we rely on the following known relationship between Berezin quantization and Weyl quantization on $C_c^\infty(\mathbb{R}^{2n})$.[68] One can show that for any $f \in C_c(\mathbb{R}^{2n})$,

$$Q_\hbar^B(f) = \int_{\mathbb{R}^{2n}} \frac{d^n a \, d^n b}{(2\pi\hbar)^n} e^{-\frac{\hbar}{4}(a \cdot a + b \cdot b)} (\mathcal{F}f)(a, b) \, \pi_S(W_\hbar(a, b)). \tag{3.21}$$

This suffices to establish that the continuous extension of the Berezin quantization map to functions in $AP(\mathbb{R}^{2n})$ takes the same functional form as in the following definition.

Example 19 (Berezin quantization on $AP(V')$) *Let V be a symplectic topological vector space with $M = V'$. Define the Poisson algebra $\mathcal{P} = \Delta(V) \subseteq AP(V')$ as the collection of finite linear combinations of the functions $W_0(f)$ on*

---

[65] One can show that $AP(V')$ is *-isomorphic to $\mathcal{W}(V, 0)$.

[66] See Binz et al. (2004b), Honegger and Rieckers (2005), and Honegger et al. (2008).

[67] See Eq. (2.111) on p. 143 of Landsman (1998a).

[68] See Eq. (2.117) on p. 144 of Landsman (1998a).

V' (see Eq. (2.35)) ranging over all f ∈ V with Poisson bracket defined as in Eq. (3.17). Define the fiber algebras by

$$\mathfrak{A}_0 = AP(V')$$
$$\mathfrak{A}_\hbar = \mathcal{W}(V, \hbar\sigma) \quad \text{for } \hbar > 0. \tag{3.22}$$

Suppose, furthermore, that V carries a complex inner product α that is compatible with σ in the sense that

$$\text{Im}(\alpha(f, g)) = \sigma(f, g) \tag{3.23}$$

for all f, g ∈ V. Then define the Berezin quantization maps $Q_\hbar^B : \Delta(V) \to \mathcal{W}(V, \hbar\sigma)$ for $\hbar > 0$ by the linear extension of

$$Q_\hbar^B(W_0(f)) := e^{-\frac{\hbar}{4}\alpha(f,f)} W_\hbar(f) \tag{3.24}$$

for all f ∈ V. Then $Q_\hbar^B$ is a strict deformation quantization.[69]

As mentioned, the central difference between Berezin quantization and Weyl quantization is that $Q_\hbar^B$ maps positive functions to positive operators, which also guarantees the continuity of $Q_\hbar^B$. This implies that $Q_\hbar^B$ has a unique continuous extension from the dense subset $\Delta(V)$ to the full domain $\mathfrak{A}_0 = AP(V')$. This is also significant for states because it implies that whenever ω is a state on $\mathcal{W}(V, \hbar\sigma)$, the linear functional $\omega \circ Q_\hbar^B$ is also a state on $AP(V')$

In the case of $V = \mathbb{R}^{2n}$, the real part of α can be taken to be the standard inner product. If V is infinite dimensional, then there will in general be different choices for the real part of α. These correspond, as we will see later, to different complex structures on V or, physically, to different "frequency-splitting" procedures of the phase space that determine positive energy modes. Although different choices for the real part of α will correspond to distinct Berezin quantization maps, they are all equivalent to the Weyl quantization maps, and hence to each other. It follows that the Berezin and Weyl quantization maps all determine the same continuous bundle of C*-algebras on the domain $AP(V')$.

These quantization procedures can be extended to the resolvent algebra as well.

Example 20 (Weyl quantization on $C_\mathcal{R}(V')$) *Let V be a symplectic topological vector space with $M = V'$. Define the Poisson algebra $\mathcal{P} = S_\mathcal{R}(V')$ as the linear span of functions g ∘ P, where P is a finite dimensional projection on V' and g is a Schwartz function on V'. One can induce a unique Poisson bracket on*

---

[69] See Honegger and Rieckers (2005) and Browning et al. (2020).

$S_{\mathcal{R}}(V')$ *from any continuous, surjective, symplectic map* $p\colon V \to \mathbb{R}^{2n}$, *and the Poisson bracket does not depend on the choice of* p. *Define the fiber algebras by*

$$\mathfrak{A}_0 = C_{\mathcal{R}}(V')$$
$$\mathfrak{A}_\hbar = \mathcal{R}(V, \hbar\sigma) \quad for\ \hbar > 0. \tag{3.25}$$

Define the Weyl quantization maps $Q_\hbar^W\colon S_{\mathcal{R}}(V') \to \mathcal{R}(V, \hbar\sigma)$ *for* $\hbar > 0$ *by*

$$Q_\hbar^W(g \circ P) := \int_{P[V']} \frac{d^n x}{(2\pi\hbar)^n}(\mathcal{F}g)(x)\,\pi_S(W_\hbar(x)) \tag{3.26}$$

for all $g \circ P \in S_{\mathcal{R}}(V')$ *with* $P[V']$ *a finite-dimensional subspace. Then* $Q_\hbar^W$ *is a strict deformation quantization.*[70]

Weyl quantization formally corresponds, via the continuous functional calculus, to the assignment[71]

$$Q_\hbar^W(R_0(\lambda, f)) = R_\hbar(\lambda, f) \tag{3.27}$$

for all $\lambda \in \mathbb{R} \setminus \{0\}$ and all $f \in V$. Note that just as for the quantization of the Weyl algebra, the quantization of the resolvent algebra is general enough to apply to infinite-dimensional phase spaces. Although we omit an extended discussion, the Berezin quantization can likewise be extended to the resolvent algebra, where it determines an equivalent quantization, and hence, the same continuous bundle of C*-algebras.

Our remaining examples show that the quantization methods discussed can be extended to physical systems with symmetries.

Example 21 (Weyl quantization for internal symmetries) *Let G be a Lie group. We might, for example, interpret elements of G as the internal symmetries of a physical system, and so corresponding to some kind of charge structure. We will assume that G is unimodular. For simplicity, we consider only the very restricted case where the exponential map is a diffeomorphism between G and the Lie algebra* $\mathfrak{g}$.[72] *One can generalize the construction to other Lie groups by working locally; we refer the reader to Landsman (1999) for the details of the more general formulation. Let* $M = \mathfrak{g}^*$ *be the dual to the Lie algebra* $\mathfrak{g}$ *of G. Define the Poisson algebra* $\mathcal{P} = C^\infty_{PW}(\mathfrak{g}^*)$ *as the collection of (Paley–Wiener) functions on* $\mathfrak{g}^*$ *whose Fourier transform is smooth and compactly supported*

---

[70] See van Nuland (2019).

[71] See van Nuland (2019, Prop. 3.4) for this characterization.

[72] The Lie algebra $\mathfrak{g}$ to a Lie group G is the tangent space at the identity $e \in G$.

with the Lie–Poisson bracket given by[73]

$$\{f,g\}(\theta) = -\theta([df(\theta), dg(\theta)]) \tag{3.28}$$

for all $f, g \in C_{PW}^\infty(\mathfrak{g}^*)$ *and* $\theta \in \mathfrak{g}^*$. *Define the fiber algebras by*

$$\mathfrak{A}_0 = C_0(\mathfrak{g}^*)$$
$$\mathfrak{A}_\hbar = C^*(G) \quad \text{for } \hbar > 0. \tag{3.29}$$

The assumption that the exponential map is a diffeomorphism allows one to identify G *with* $\mathfrak{g}$ *via the association* $X \in \mathfrak{g} \mapsto \exp X \in G$. *We can use this to specify an element of* $C^*(G)$ *as a function on* G *by its values on elements* $\exp X \in G$, *as follows. Define the Weyl quantization maps* $Q_\hbar^W : C_{PW}^\infty(\mathfrak{g}^*) \to C^*(G)$ *for* $\hbar > 0$ *by*

$$Q_\hbar(f)(\exp X) := \int_{\mathfrak{g}^*} \frac{d^n\theta}{(2\pi\hbar)^n} f(\theta) e^{\frac{i\theta(X)}{\hbar}} \tag{3.30}$$

for all $f \in C_{PW}^\infty(\mathfrak{g}^*)$ *and* $X \in \mathfrak{g}$. *Then* $Q_\hbar^W$ *is a strict deformation quantization.*[74]

Example 22 (Weyl quantization for external symmetries) *Let* G *be a Lie group acting on a manifold* Q. *Usually, we interpret* Q *as a configuration space and elements of* G *as external symmetries of a physical system. We again assume that* G *is unimodular, and we again consider only the very restricted case where the exponential map is a diffeomorphism between* G *and* $\mathfrak{g}$ *and refer the reader to Landsman (1999) for the more general construction. Let* $M = \mathfrak{g}^* \times Q$ *and define the Poisson algebra* $\mathcal{P} = C_{PW}^\infty(\mathfrak{g}^* \times Q)$ *as the collection of functions whose fiberwise Fourier transform is smooth and compactly supported with a Poisson bracket given by a generalization of Eq. (3.28) (See Landsman 1998, Eq. (3.101), p. 297). Define the fiber algebras by*

$$\mathfrak{A}_0 = C_0(\mathfrak{g}^* \times Q)$$
$$\mathfrak{A}_\hbar = C^*(G, Q) \quad \text{for } \hbar > 0. \tag{3.31}$$

We again identify $X \in \mathfrak{g}$ *with* $\exp X \in G$ *and define the Weyl quantization maps* $Q_\hbar^W : C_{PW}^\infty(\mathfrak{g}^*) \to C^*(G, Q)$ *for* $\hbar > 0$ *by*

$$Q_\hbar^W(f)(\exp X, q) := \int_{\mathfrak{g}^*} \frac{d^n\theta}{(2\pi\hbar)^n} f(\theta, \exp(-\frac{1}{2}X)q) e^{\frac{i\theta(X)}{\hbar}} \tag{3.32}$$

for all $f \in C_{PW}^\infty(\mathfrak{g}^*)$, $X \in \mathfrak{g}$, *and* $q \in Q$. *Then* $Q_\hbar^W$ *is a strict deformation quantization.*[75]

---

[73] Here, d denotes the exterior derivative on $\mathfrak{g}^*$, and we use the fact that the cotangent space at any $\theta \in \mathfrak{g}^*$ is canonically isomorphic to $\mathfrak{g}$ to understand the Lie bracket $[\cdot, \cdot]$.

[74] See Rieffel (1993), Landsman (1998b, 1999), and chapter 7.2 of Landsman (2017).

[75] See Landsman (1990a,b, 1999) and chapter 7.3 of Landsman (2017).

Example 23 (Weyl quantization for a particle in an external gauge field) *Let* $P \to Q$ *be a principal bundle over a base space given by a Riemannian manifold* Q *with fiber given by a Lie group* G*. One can use such a structure to represent a particle moving in the configuration space* Q *subject to external forces from a gauge field represented by a principal connection on* P *with gauge symmetry group* G*. After taking into account the symmetries of the system via Marsden–Weinstein reduction, the reduced Hamiltonian dynamics can be represented on the so-called universal phase space* $M = (T^*P)/G$*, where* G *carries the natural action on* T^*P *by the pullback. Notice that* $C_0((T^*P)/G) \cong C_0(T^*P)^G$*, where we use the superscript* G *to denote the collection of* G*-invariant functions. Thus, we can consider quantizing functions on* T^*P *directly. Define the Poisson algebra* $\mathcal{P} = C^\infty_{PW}(T^*P)^G$ *as the collection of* G*-invariant functions on* (T^*P) *whose fiberwise Fourier transform is smooth and compactly supported with the standard Poisson bracket on* T^*P*. We again assume* G *is unimodular, and we now assume* G *is compact. Finally, we assume* P *carries a* G*-invariant measure. Define the fiber algebras by*

$$\mathfrak{A}_0 = C_0(T^*P)^G$$
$$\mathfrak{A}_\hbar = \mathcal{K}(L^2(P))^G \quad \text{for } \hbar > 0. \tag{3.33}$$

Here, we also use the superscript G *to denote the* G*-invariant compact operators under the natural action of* G *by unitary conjugation (see Eq. (6.6)). Define the Weyl quantization maps* $Q^W_\hbar : C^\infty_{PW}(T^*P)^G \to \mathcal{K}(L^2(P))^G$ *for* $\hbar > 0$ *by*

$$(Q^W_\hbar(f)\psi)(p) := \int_P (k_\hbar(f))(p,p')\psi(p')dp' \tag{3.34}$$

for $f \in C^\infty_{PW}(T^*P)^G$*,* $\psi \in L^2(P)$*, and* $p \in P$*, with kernel*

$$k_\hbar(f)(\exp_x X/2, \exp_x -X/2) := \hbar^{-n}(\mathcal{F}f)(X) \tag{3.35}$$

for $X \in T^*_x P$ *at suitably small* $\hbar$ *and in suitably local regions of* $P \times P$*. The kernel* $k_\hbar(f)$ *is defined to be zero when the appropriate conditions do not hold. Then* $Q^W_\hbar$ *is a strict quantization. It follows that* $\mathcal{K}(L^2(P))^G \cong \mathcal{K}(L^2(Q)) \otimes C^*(G)$*.*[76]

We mention that the previous quantization procedures for symmetries in Examples 21–23 can be unified as special cases of quantization for Lie groupoids, where the phase space is the dual to the associated Lie algebroid

---

[76] See Landsman (1993) and Landsman (1999). Note that Eqs. (3.34)–(3.35) can be applied to arbitrary $f \in C^\infty_{PW}(T^*P)$ – not only the G-invariant functions. We will make use of this fact later in Section 6.

and the quantized algebra is the (reduced) groupoid C*-algebra.[77] We will, however, not use this level of generality in what follows.

3.3 Other Approaches to Quantization

3.3.1 Mackey's Systems of Imprimitivity

Mackey developed an approach to quantization following Weyl's program for incorporating symmetries into quantum theories.[78] The central idea is that the canonical commutation relations are reformulated in terms of the action of a symmetry group. This in turn is captured by the result of Weyl quantization for systems with symmetries.

We first present Mackey's approach through a simple (but illustrative) example.

Example 24 (Translation symmetry on $\mathbb{R}^n$) *Consider a physical system consisting of point particles moving in the configuration space $Q = \mathbb{R}^n$. To quantize this theory, we consider the canonical commutation relations for position and momentum*

$$[Q_j, P_k] = i\hbar\delta_{jk} \tag{3.36}$$

for $j, k = 1, \ldots, n$ corresponding to the canonical coordinates $(q_1, \ldots, q_n, p_1, \ldots, p_n)$ on the phase space T^*Q. Of course, one can represent these relations on $L^2(\mathbb{R}^n)$ with the standard position and momentum operators. In this case, one can show that each P_k is the generator of a translation symmetry group for Q_k in the sense that the one-parameter unitary group $U_k(a) := e^{iaP_k}$ for $a \in \mathbb{R}$ satisfies

$$U_k(a)Q_k U_k(-a) = Q_k - aI \tag{3.37}$$

for any $a \in \mathbb{R}$. Putting this all together, one has a strongly continuous unitary representation of the entire translation group $G = \mathbb{R}^n$ by

$$x = (a_1, \ldots, a_n) \mapsto U(x) := U_1(a_1) \ldots U_n(a_n) \tag{3.38}$$

for $x \in G$. Conversely, one can suppose only that one starts with a unitary representation $U(x)$ for $x \in G$ satisfying the translation symmetry condition given in Eq. (3.37). If the map $x \mapsto U(x)$ is strongly continuous, then Stone's theorem guarantees that each one-parameter family $U_k(a)$ has a self-adjoint generator P_k satisfying the canonical commutation relations for Q_k and P_k.

---

[77] See Landsman (1999) for this unification in terms of Lie groupoids. See also Bieliavsky and Gayral (2015) for a broad class of generalizations of Weyl quantization.

[78] See Mackey (1968), Landsman (2007), and chapter 7.4 of Landsman (2017).

The use of unitary operators to represent the translation symmetries circumvents reference to the unbounded momentum operator. One can also avoid reference to the unbounded position operator by considering instead bounded functions of Q_k, in particular restricting attention to representations of the C*-algebra $C_0(Q)$. Since the translation group $G = \mathbb{R}^n$ acts on $Q = \mathbb{R}^n$ by

$$(x, q) \in G \times Q \mapsto q + x \in Q, \tag{3.39}$$

we also have an action L_x of G on $C_0(Q)$ by

$$(L_x f)(q) = f(q - x) \tag{3.40}$$

for $f \in C_0(Q)$, $q \in Q$, and $x \in G$. Then the canonical commutation relations are encoded in the following relation for a unitary representation U of G and a *-representation π of $C_0(Q)$:

$$U(x)\pi(f)U(x) = \pi(L_x f) \tag{3.41}$$

for $f \in C_0(Q)$ and $x \in G$.

All one needs to formulate the condition in Eq. (3.41) is a configuration space carrying the action of a symmetry group. It is only accidental in this case that the configuration space $Q = \mathbb{R}^n$ and the translation symmetry group $G = \mathbb{R}^n$ are the same. One can generalize this idea to other configuration spaces Q that are smooth manifolds carrying the action of a Lie group G. In this case, one keeps Eq. (3.41) and only replaces the definition of L_x by

$$(L_x f)(q) = f(x^{-1}q) \tag{3.42}$$

for $f \in C_0(Q)$, $q \in Q$, and $x \in G$, where we allow any smooth action of the group G on Q. The pair consisting of a strongly continuous unitary representation of G with a faithful representation of $C_0(Q)$ is called a *system of imprimitivity*.

Much of Mackey's work focused on finding and classifying irreducible systems of imprimitivity – that is, systems of imprimitivity (U, π) represented on a Hilbert space $\mathcal{H}$ containing no subspaces invariant under the joint actions of all operators $U(x)$ for $x \in G$ and $\pi(f)$ for $f \in C_0(Q)$. Wigner's well-known principle asserts that each such irreducible representation should correspond to a different kind of fundamental particle with generalized charge associated with the symmetry group G. Since the strongly continuous irreducible unitary representations of G are in bijective correspondence with the irreducible representations of $C^*(G)$, it follows that the irreducible systems of imprimitivity are in bijective correspondence with the irreducible unitary representations of $C^*(G, Q)$. Hence, Weyl quantization of the action of G on Q coincides with quantization by a system of imprimitivity. In other words, there is a sense in which Mackey's approach is a special case of strict deformation quantization – specifically, the strict deformation quantization of systems with symmetry.

The central result in this tradition, known as the *imprimitivity theorem*, classifies the irreducible systems of imprimitivity for a suitable action of a group G on a manifold Q – or, equivalently, the irreducible representations of $C^*(G, Q)$ – as in bijective correspondence with pairs of G-orbits in Q and unitary representations of the so-called stabilizer group of a point in Q.

3.3.2 Segal's Field Quantization

There is a well known procedure for constructing bosonic quantum field theories on symmetric Fock spaces through the Segal quantization method.[79] This section shows a sense in which the Segal method leads to the same results as the Weyl quantization procedure on infinite-dimensional phase spaces.

Consider a symplectic topological vector space V with symplectic form σ. As previously, V is typically taken to represent the phase space for a linear field theory, but the procedure works just as well for finite-dimensional V. The Segal method is to search for a complex inner product α on V compatible with the symplectic form on V. When V is a real vector space, this can be done by choosing a *complex structure*. A complex structure is a linear operator J on V satisfying $J^2 = -I$, so that it can be understood as multiplication by the complex unit, thereby turning the real vector space V into a complex vector space with

$$if := Jf \tag{3.43}$$

for all $f \in V$. The complex structure J is compatible with the symplectic form σ if, for all $f, g \in V$, we have

$$\sigma(f, Jf) \geq 0 \qquad\qquad \sigma(Jf, Jg) = \sigma(f, g). \tag{3.44}$$

Given a compatible complex structure, one can define a complex inner product α on V by

$$\alpha(f, g) := \sigma(f, Jg) + i\sigma(f, g) \tag{3.45}$$

for all $f, g \in V$. Next, one constructs the completion of V with respect to α to form a Hilbert space $\mathcal{H} = \overline{V}^{\alpha}$. This allows us to construct the symmetric (bosonic) Fock space

---

[79] See Segal (1963), Weinless (1969), Kay (1979), Baez et al. (1992), or Clifton and Halvorson (2001). The procedure also applies to fermionic quantum field theories through the construction of antisymmetric Fock spaces.

$$\mathcal{F}(\mathcal{H}) := \bigoplus_{n=0}^{\infty} \mathcal{H}_n \qquad\qquad \mathcal{H}_n := \mathcal{S}\left(\bigotimes_{k=0}^{n} \mathcal{H}\right), \qquad (3.46)$$

where $\mathcal{S}(\bigotimes_{k=0}^{n} \mathcal{H})$ denotes the symmetric subspace of $\bigotimes_{k=0}^{n} \mathcal{H}$, with the stipulation that $\mathcal{H}_0 = \mathbb{C}$. Such Fock spaces provide the standard setting for free quantum field theory. One can define unbounded self-adjoint (smeared) field operators $\phi(f)$ for each $f \in V$ on this space, as follows. Take the domain

$$\mathcal{D} := \left\{ \sqrt{\exp(\psi)} := \sum_{k=0}^{\infty} \frac{\otimes^k \psi}{\sqrt{k!}} \,\Big|\, \psi \in \mathcal{H} \right\} \qquad (3.47)$$

and define $\phi(f)$, for any $f \in V$, by

$$\phi(f)\sqrt{\exp(\psi)} := i\left(\alpha(f,\psi)\sqrt{\hbar} \cdot \sqrt{\exp(\psi)} - \sqrt{\hbar} \cdot \frac{d}{dt}\Big|_{t=0} \sqrt{\exp(\psi + tf)} \right) \quad (3.48)$$

for any $\psi \in \mathcal{H}$. Working on the Hilbert space $\mathcal{F}(\mathcal{H})$ with the operators $\phi(f)$ recovers standard free quantum field theory, as follows.

One can define, for each $f \in V$, the (smeared) creation and annihilation operators

$$a^*(f) := \frac{1}{\sqrt{2}}(\phi(f) - i\phi(Jf)) \qquad a(f) := \frac{1}{\sqrt{2}}(\phi(f) + i\phi(Jf)) \qquad (3.49)$$

satisfying the canonical commutation relations, and the (smeared) number operator

$$N(f) := a^*(f)a(f). \qquad (3.50)$$

Finally, for an α-orthonormal basis $\{f_k\}$ of V, define the total number operator by

$$N := \sum_{k=1}^{\infty} N(f_k). \qquad (3.51)$$

Each subspace $\mathcal{H}_n \subseteq \mathcal{F}(\mathcal{H})$ consists of eigenvectors of N with eigenvalue n, so that vectors in $\mathcal{H}_n$ are interpreted as states with n particles. Moreover, $a^*(f)$ raises the number of particles by mapping vectors in $\mathcal{H}_n$ to vectors in $\mathcal{H}_{n+1}$ while $a(f)$ lowers the number of particles by mapping vectors in $\mathcal{H}_{n+1}$ to vectors in $\mathcal{H}_n$. The vacuum state is the (unique up to a phase) vector Ω_0 with

$$a(f)\Omega_0 = 0 \qquad\qquad \langle \Omega_0, N(f)\Omega_0 \rangle = 0 \qquad (3.52)$$

for all $f \in V$, or in other words $\Omega_0 = 1 \in \mathcal{H}_0 \subseteq \mathcal{F}(\mathcal{H})$.

Example 25 (Klein–Gordon field) *Consider a scalar field φ on Minkowski spacetime $\mathbb{R}^4$ satisfying the Klein–Gordon equation*

$$\frac{\partial^2}{\partial t^2}\varphi - \nabla^2\varphi = -m^2\varphi \tag{3.53}$$

for some constant $m > 0$. We consider the space of initial data (π, φ), where π is the canonically conjugate momentum to φ. Let $V = C_c^\infty(\mathbb{R}^3) \oplus C_c^\infty(\mathbb{R}^3)$ be the space of real-valued test functions for initial data on a spacelike hypersurface with the symplectic form

$$\sigma((f,g),(\tilde{f},\tilde{g})) := \int_{\mathbb{R}^3} f(x)\tilde{g}(x) - \tilde{f}(x)g(x)\, d^3x \tag{3.54}$$

for all $(f,g),(\tilde{f},\tilde{g}) \in V$. We now define a compatible complex inner product α on V through a complex structure that commutes with the time evolution of the Klein–Gordon equation. The complex structure J is a linear operator on V defined by

$$J(f,g) = (-\mu^{-1}g, \mu f) \qquad\qquad \mu = (m^2 - \nabla^2)^{1/2}, \tag{3.55}$$

for all $(f,g) \in V$, where μ is a positive operator on $C_c^\infty(\mathbb{R}^3)$. With $\mathcal{H} = \overline{V}^\alpha$ the completion of V relative to the inner product α, the symmetric Fock space $\mathcal{F}(\mathcal{H})$ is the standard Hilbert space for the mass m Klein–Gordon field in inertial reference frames with field operators $\phi(f)$. The Fock space $\mathcal{F}(\mathcal{H})$ carries a unitary representation of the Poincaré group with the Klein–Gordon dynamics as time evolution.

How does the Segal method relate to Weyl quantization? Consider the quantization of the Weyl algebra $Q_\hbar: AP(V') \to \mathcal{W}(V, \hbar\sigma)$. Define a state ω_0 on $\mathcal{W}(V, \hbar\sigma)$ by the continuous linear extension of

$$\omega_0(W_\hbar(f)) := e^{-\frac{\hbar}{4}\alpha(f,f)} \tag{3.56}$$

for all $f \in V$. One can show that the GNS representation π_{ω_0} for the state ω_0 has the Hilbert space

$$\mathcal{H}_{\omega_0} = \mathcal{F}(\mathcal{H}) \tag{3.57}$$

and π_{ω_0} takes the form

$$\pi_{\omega_0}(W_\hbar(f)) = e^{i\phi(\sqrt{\hbar}\cdot f)} \tag{3.58}$$

for all $f \in V$, with the GNS vector given by the vaccum vector

$$\Omega_{\omega_0} = \Omega_0. \tag{3.59}$$

Thus, the bosonic Fock space constructed from the inner product α can also be obtained as the GNS representation of the abstract vacuum state determined by α on the Weyl algebra.

3.3.3 Formal Deformation Quantization

Strict quantization treats the classical–quantum correspondence via the $\hbar \to 0$ limit, understood as a genuine limiting relation where $\hbar$ takes on numerical values and the algebraic relations between physical quantities vary in an appropriately continuous way between a classical and a quantum theory. A different approach to the same limit understands small values of Planck's constant by treating $\hbar$ as a formal parameter in asymptotic power series expansions of physical quantities. This approach is called formal deformation quantization.[80]

Recall that a strict quantization with quantization maps $Q_\hbar : \mathcal{P} \to \mathfrak{A}_\hbar$ is called a deformation quantization when $Q_\hbar$ is faithful and $Q_\hbar[\mathcal{P}]$ is closed under algebraic operations. This allows one to pull back the noncommutative multiplication operation from $\mathfrak{A}_\hbar$ and define a noncommutative product on the Poisson algebra of classical functions one began with. Formal deformation quantization directly defines noncommutative products on functions on a phase space, where both the functions and the product are understood in terms of formal power series. In what follows, $C^\infty(\mathcal{M})[[\hbar]]$ is the space of formal power series in a formal parameter $\hbar$ with coefficients in the smooth functions on a Poisson manifold $\mathcal{M}$. Again, $\mathcal{M}$ typically represents the phase space of a physical system.

Definition 7 A *formal ⋆-product* on $C^\infty(\mathcal{M})[[\hbar]]$ is a $\mathbb{C}[[\hbar]]$ bilinear map

$$\star : C^\infty(\mathcal{M})[[\hbar]] \times C^\infty(\mathcal{M})[[\hbar]] \to C^\infty(\mathcal{M})[[\hbar]] \tag{3.60}$$

of the form

$$f \star g = \sum_{n=0}^{\infty} C_n(f,g)\hbar^n, \tag{3.61}$$

where for each $n \in \mathbb{N}$, $C_n : C^\infty(\mathcal{M}) \times C^\infty(\mathcal{M}) \to C^\infty(\mathcal{M})$ is a $\mathbb{C}[[\hbar]]$-bilinear map. One requires $\star$ to satisfy for all $f, g, h \in C^\infty(\mathcal{M})[[\hbar]]$:

(i) $f \star (g \star h) = (f \star g) \star h$;
(ii) $C_0(f,g) = fg$;

---

[80] See Bayen et al. (1978a,b) for the origin of this approach, as well as Bordemann and Waldmann (1998) and Waldmann (2005, 2016) for a sampling of recent work. Historically, formal deformation quantization preceded and provided the inspiration for the approach of strict quantization that is our focus here.

(iii) $C_1(f,g) - C_1(g,f) = i\{f,g\}$;

(iv) $1 \star f = f \star 1 = f$;

(v) for each $n \in \mathbb{N}$, C_n is a bidifferential operator in the sense that it is a finite linear combination of derivative operators acting on each of its arguments.

In a formal $\star$-product, the operators C_n should be understood as different orders of approximation in $\hbar$ of the product. The first condition requiring associativity matches exactly the condition of associativity on C*-algebras. The second condition says that the zeroth order C_0 of the formal $\star$-product should match the classical product of functions, thus corresponding to von Neumann's condition in a strict quantization. The third condition says that the first order C_1 of the formal $\star$-product should match the Poisson bracket multiplied by the imaginary unit, thus corresponding to Dirac's condition in a strict quantization. The fourth condition says that the constant function 1 always serves as a unit of the algebra of formal power series relative to the $\star$-product. The fifth condition is a technical condition satisfied by all known examples.

Just as with strict deformation quantization, a number of examples of formal deformation quantization have been explicitly constructed.[81] However, the existence of formal deformation quantizations of general Poisson manifolds has also been proven, and these formal deformation quantizations have been completely classified mathematically. This is the content of the celebrated Kontsevich theorem (Kontsevich, 2003).

To relate formal deformation quantization to strict deformation quantization, one can attempt to show that the expression of a formal $\star$-product converges for some suitable class of functions with convergent power series expansions. Recent work has shown this to be possible for a number of examples, including Weyl quantization on linear spaces in both finite and infinite dimensions, as well as for Weyl quantization of the dual to a Lie algebra (Waldmann, 2019). We expect that when a formal $\star$-product converges, it yields a strict deformation quantization.

4 Intertheoretic Reduction

Section 3 provided an overview of the mathematical foundations of quantization and the $\hbar \to 0$ limit. In this section, we begin in earnest our discussion of the philosophical significance of those mathematical tools. The first

---

[81] See, for example, Gutt (1983) for formal deformation quantization of the dual to a Lie algebra, and see Fedosov (1996) for formal deformation quantization of symplectic manifolds.

philosophical topic we consider is the possibility of an intertheoretic reduction between classical and quantum mechanics. While we will avoid wading too deeply into the general philosophical literature on the nature of intertheoretic reduction, we will at least circumscribe the issue so that we can discuss some of the particular controversies with the $\hbar \to 0$ limit.

We will take for granted that an intertheoretic reduction, if one is possible, would be constituted by an explanation of the success of classical mechanics on the basis of quantum mechanics. Some authors, going back to the canonical works of Nagel (1961, 1998), have thought that such an explanation would be required to take the form of a deductive argument in some logical language. On this scheme, the premises would consist in laws from the more fundamental quantum theory, supplemented by bridge principles that connect the distinctive concepts of the lower-level quantum theory to those of the higher-level classical theory. However, Nickles (1973) made prominent the idea that important intertheoretic relations could be captured by limits of parameters, and while he suggests such limits often have a heuristic role different from the kinds of explanations Nagel sought, Nickles also suggests that a limiting relation could be used to explain why a theory's predecessor was successful in the regimes it was. This is exactly the structure of the explanation the $\hbar \to 0$ limit hopes to provide – an explanation of why classical physics is successful in certain regimes on the basis of the structure of quantum theory.

Questions abound concerning the nature of such an explanation of the success of classical physics. First, what exactly are we aiming to explain? The explanandum has been characterized as the success of classical physics, but much more needs to be said about what that success consists in. Second, does the $\hbar \to 0$ limit have the required mathematical properties to ensure that it can produce a theory of classical physics on the basis of quantum physics alone? The technical status of the $\hbar \to 0$ for this task has remained controversial due to purported "singularities" in the limit. Third, how precisely does the limit $\hbar \to 0$ provide an explanation? A common objection has been that since $\hbar$ is a constant of nature, it could only take different values in distinct possible worlds, and the nature of those other worlds does not bear on the success of classical physics in our own world. So is it possible to use the mathematical tools of quantization and the $\hbar \to 0$ limit to provide a reductive explanation? We take up each of these questions in the following sections in an attempt to clarify the extent to which one can explain classical physics from quantum theory.

4.1 Empirical vs. Theoretical Reduction

First, we take up the task of characterizing the explanandum of a reduction of classical to quantum mechanics. What does it mean to explain the success of classical physics? We shall focus on two contrasting answers:[82]

- *Empirical Reduction*: An explanation of the success of classical physics is an explanation of why the *empirical predictions* of classical physics are approximately correct in certain regimes.
- *Theoretical Reduction*: An explanation of the success of classical physics is an explanation of why the *theoretical structure* of classical physics is appropriate for use in certain regimes.

As we will see, theoretical reduction is strictly stronger than empirical reduction. However, each of these two characterizations of the explanandum is still vague as it stands. We aim to provide some clarification in what follows.

First, what are the empirical predictions of classical physics that are the target of explanation for an empirical reduction? We take these empirical predictions to be expectation values or probability values for the results of certain measurements in certain states. However, these predictions split into two broad classes. First, and less often mentioned, classical physics makes predictions about the possible correlations allowed between multiple measurements of a physical system. That is, classical physics does not allow for the Bell-inequality violating correlations produced in entangled quantum states. So one explanatory question is: why – if a physical system is better represented by a quantum state that *does* produce characteristically quantum correlations – can one (under some circumstances) approximately recover predictions of expectation values and probability values from a classical state that does not allow for such correlations? Second, classical physics predicts that the expectation values for physical quantities will evolve dynamically and change according to, for example, Hamiltonian equations of motion that differ from the central dynamical equations of quantum mechanics. So another explanatory question is: why – if the temporal evolution of a physical system is better represented by quantum dynamics – can one (under some circumstances) approximately recover predictions for how expectation values and probability values change in time from classical dynamics?

---

[82] Here, we adapt the helpful terminology employed by Rosaler (2015a). Our notion of *theoretical reduction* does not exactly line up with Rosaler's notion of *formal reduction*, however, because the latter emphasizes formal aspects of the mathematics used in physical theories. As we have seen in Section 3.3.3, the term "formal" has a precise mathematical meaning in the context of deformation quantization, while all of the tools we consider in the remainder of our investigation are nonformal.

Ultimately, we will aim to answer these explanatory questions indirectly by first employing strict quantization to explain the theoretical structure of classical physics in a theoretical reduction. Then we will establish that the explanation of the theoretical structure of classical physics offered implies an explanation of the empirical predictions of classical physics, and so provides an empirical reduction. However, if one thought a theoretical reduction was not possible, one might aim to directly provide an empirical reduction.

Let us take a moment to illustrate how one might try to answer these explanatory questions of an empirical reduction on their own by considering the work of Rosaler (2015a, 2016, 2018). Rosaler explains the appearance of approximately classical correlations by invoking the well-known mechanism of decoherence. He argues that when a quantum system interacts with its environment, the joint dynamical evolution leads (typically on very short time scales) to quantum states that are localized around certain "pointer states." If the interaction with the environment constitutes a kind of measurement of the quantum system, then these pointer states are associated with possible measured values for the quantity being measured. For example, for an n-particle system, one might take the pointer states to be the coherent states, whose distributions for position and momentum are sharply peaked around points in the classical phase space. The idea is that the correlations exhibited by pointer states are approximately classical by yielding expectation values for observables close to those of the classical state they are centered on, and the decoherence process explains how it is that these quasi-classical states arise dynamically.[83] Rosaler explains the appearance of classical dynamics then by showing that a version of Ehrenfest's theorem, generalized to open quantum systems, implies that the expectation values of a system interacting with its environment will stay close to the determinate values of a classical state whose initial conditions agree with the initial expectations of the quantum system and then evolve in time according to classical equations of motion.

Let us see how Rosaler makes his approximation condition precise by considering again an n-particle system. We provide here only some very basic results to give a flavor of the approximations involved; there are many more results concerning semiclassical approximations to quantum dynamics.

---

[83] One of Rosaler's aims in employing decoherence theory is to avoid committment to any particular interpretation of quantum theory, or solution to the measurement problem. Many interpretations of quantum theory purport to explain the appearance of quasi-classical behavior at least in measurement setups by changing the structure of quantum theory. But one might wonder (as Rosaler does, and as we do here) whether it is possible to explain quasi-classical behavior on the basis of only the structure of the standard quantum theory, which all solutions to the measurement problem in some sense share or vindicate. See also Rosaler (2015b, 2016) and Romano (2016).

Suppose the system starts in a quantum state given by a wavefunction ψ whose evolution by the Schrödinger equation leads to a time evolving state $\psi(t)$. Similarly, one can consider the classical state given by a phase space point with $q_j := \langle \psi, Q_j \psi \rangle$ and $p_j := \langle \psi, P_j \psi \rangle$. One can represent the time evolving classical state by the trajectories $q_j(t)$ and $p_j(t)$ obeying the classical equations of motion. Rosaler shows that for certain position and momentum measurement error bounds ϵ_Q and ϵ_P, there is some timescale τ such that

$$|\langle \psi(t), Q_j \psi(t) \rangle - q_j(t)| < \epsilon_Q$$
$$|\langle \psi(t), P_j \psi(t) \rangle - p_j(t)| < \epsilon_P \tag{4.1}$$

whenever $t < \tau$. The error bounds and appropriate timescale here may depend on both the dynamics and the particular state one begins with. But Rosaler takes such results to justify claims of an empirical reduction between classical and quantum mechanics.

However, one might wonder if such an empirical reduction is all that can be achieved. The question of whether one can provide a theoretical reduction between classical and quantum mechanics is the question of whether one can explain *more* than the empirical predictions of classical physics. After all, classical physics is more than just a list of empirical predictions; it involves a theoretical structure built from phase spaces with Poisson brackets and Hamiltonian equations of motion. And it is an empirical fact that users of the classical theory employing this theoretical structure have had success in producing predictions that match experiments in certain regimes. Why is it that users of classical theoretical structures have been successful in this way?

How one goes about answering this question depends on how one understands classical physics. For example, even Nagel (1998, p. 911) noticed that if one takes a particular kind of instrumentalist attitude toward classical physics by understanding the theory as equivalent to its empirical predictions, then all one needs to explain is the predictions it gives rise to. On this instrumentalist attitude, then, theoretical reduction is no more than empirical reduction. However, if one believes that there is content to classical physics not captured in its empirical predictions, or if one thinks that classical physics is not equivalent to its empirical predictions, then there is more to be explained.[84]

---

[84] For other authors who have characterized the reduction of classical to quantum physics as requiring more than an explanation of the empirical success of classical physics, see Post (1971), Radder (1991), and Feintzeig (2020). See also Scheibe (1986, 1999), Rohrlich (1990), and Primas (1998) for more work in this tradition.

One can organize the theoretical structure of classical physics into at least the following three parts.[85] These are related to various layers of mathematical structure in models of classical physics. We list them here as corresponding to explanatory why-questions that a theoretical reduction aims to answer.

1. Why are the classical physical quantities so structured as to be described by functions on a phase space? And why are classical states so structured as to be described by points in a phase space, or probability measures thereon?
2. Why do classical physical quantities possess the structure of a Poisson bracket?
3. Why do classical physical quantities obey Hamiltonian dynamics?

To answer the first question, one needs to show why the bounded classical physical quantities have the algebraic structure of a commutative C*-algebra, whose pure state space (Gelfand dual) consists in the classical phase space. To answer the second question, one needs to show why that C*-algebra carries a (densely defined) Poisson bracket. To answer the third question, one needs to show why the Hamiltonian dynamics – determined through the Poisson bracket – agree with the dynamics of the quantum theory – determined by unitary evolution.

One way to distinguish the kinds of explanations sought in theoretical reduction from those sought in empirical reduction is that the latter might be state-dependent while the former should not. Recall that Rosaler's empirical reduction established that for a given state, there are error bounds and time-scales such that the classical predictions for that state are within the error bounds of the quantum predictions for those timescales. But it should be no surprise that there is *always* some large enough error bound and some short enough timescale for which such a statement holds true. For this, one need not actually employ the theoretical structure of classical or quantum physics in its entirety; this relies only on the continuity of trajectories of expectation values in each theory. What would be more substantive is if one could establish the stronger claim that there is a *fixed error bound* such that the classical predictions lie within that error bound from the quantum predictions *no matter what state one began with*. Finding such a uniform approximation across all states would indeed show that the theoretical structure of the physical quantities of

---

[85] Landsman and Reuvers (2013) and Landsman (2013, 2017) add another question to this list, which they argue is closely related to the measurement problem: why do quantum systems end up in states approximating classical pure states following measurement-like interactions? Many of the naïve ways of taking classical limits of the quantum states that result after measurement-like interactions actually lead to classical mixed states and hence do not explain why definite measurement outcomes occur. This is, of course, a central foundational issue in the relationship between classical and quantum physics. However, we leave it aside in our discussion that follows.

classical physics is appropriate for use in certain regimes. It would show that the structure of the algebra of physical quantities in a quantum theory approximates that of the classical theory, the structure of the state space of a quantum theory approximates that of the classical theory, and the structure of the dynamics of the quantum theory approximates that of the classical theory. Such a result would be much stronger than the mere approximation of classical expectation values from quantum expectation values for each state separately. As we shall see, such uniform approximations are provided by strict quantization in the $\hbar \to 0$ limit.

In the remainder of this section, we investigate the extent to which it is possible to provide a theoretical reduction between classical and quantum physics. Clearly, if a theoretical reduction can be found, then it would imply an empirical reduction as well – because the theoretical structures of classical physics give rise to the empirical predictions that are approximately accurate. Still, the possibility of such a reduction through the $\hbar \to 0$ limit has been contentious, and so we will attend to the controversies besetting it in what follows.

4.2 Singularities

It has been claimed by a number of authors that the classical $\hbar \to 0$ limit is plagued by mathematical difficulties. For example, Berry (1994), Batterman (2002), and Bokulich (2008) all claim that the classical limit is "singular" in some way that would preclude a reductive explanation of classical physics. It will be helpful to start by considering a motivating example for Batterman. He considers a family of wavefunctions solving the Schrödinger equation and attempts to take their $\hbar \to 0$ limit. Batterman employs the WKB approximation, a common tool for recovering semiclassical behavior, by using wavefunctions of the form $\psi(x) = ae^{iS(x)/\hbar}$ for some real-valued function S on a configuration space and complex constant a. Such functions do not converge in the limit as $\hbar \to 0$ even pointwise, and so Batterman claims the lack of an $\hbar \to 0$ limit in this case is an instance of a failure to reproduce classical behavior.

But the issue is also deeper than this failure of convergence. The more general problem for reduction is that there are instances where information or structure from classical physics appears to play an essential role in the derivation or explanation of new theoretical phenomena in the semiclassical regime. Batterman gives an example in chaotic systems, whose classical dynamics is difficult to recover mathematically from quantum mechanics using the geometrical techniques that Batterman favors for the $\hbar \to 0$ limit. Even still, the energy spectrum associated with the dynamics of such systems can often be determined through a "semi-classical trace formula" (Batterman, 2002, p. 110).

These results are involved and would take us too far afield for current purposes, but it suffices to notice that the derivation of these semiclassical trace formulae involve knowledge of the *classical* solutions to the equations of motion. The semiclassical physics captured in such a trace formula is, according to Batterman, not determined by the quantum theory alone, but only with the essential aid of the corresponding classical theory. And so, Batterman claims such phenomena block the reduction of classical to quantum physics because the quantum theories do not possess sufficient information to explain the semiclassical phenomena on their own. Instead, Batterman thinks of classical and quantum physics as providing structures that are in some ways independent and work together to produce explanations of semiclassical behavior.[86]

Although it may be distinct from the central concerns of those authors that have claimed the $\hbar \to 0$ limit is singular in an important sense, this suggests a more general question. Can the $\hbar \to 0$ limit be understood to determine structures of classical physics from only the structure of quantum theories? In what follows, we will outline a sense in which the answer is "yes." This answer, due to Steeger and Feintzeig (2021a,b) demonstrates a precise mathematical sense in which the theoretical structure of a classical theory at $\hbar = 0$ can be determined from the structure of a quantum theory at $\hbar > 0$.

To see this, let us consider in turn each of the layers of mathematical structure outlined in the previous section: quantities, Poisson bracket, and dynamics. We begin with the algebraic structure of physical quantities. Suppose we want to recover the structure of a commutative C*-algebra $\mathfrak{A}_0$ of functions on a classical phase space from a quantum theory representing physical quantities by elements of a noncommutative C*-algebra $\mathfrak{A}_\hbar$. Of course, a continuous bundle of C*-algebras $((\mathfrak{A}_\hbar, \phi_\hbar)_{\hbar \in [0,1]}, \mathfrak{A})$ over the interval $[0, 1]$ would carry enough information to determine the limit algebra. However, such a mathematical object clearly already contains information about the classical theory encoded in the fiber algebra at the value $\hbar = 0$. So instead, we will assume that we start with only the information from this bundle belonging to the quantum theory at $\hbar > 0$. That is, we will assume we have a family of algebras obtained from a continuous bundle of C*-algebras by "forgetting" the classical algebra at $\hbar = 0$.

To do so, we keep the fiber algebras $(\mathfrak{A}_\hbar)_{\hbar \in (0,1]}$ at $\hbar > 0$, but we consider the C*-algebra $\tilde{\mathfrak{A}}$ of continuous sections in $\mathfrak{A}$ restricted to the smaller domain $(0, 1]$. Explicitly, we define $\tilde{\mathfrak{A}}$ as the collection of elements $\tilde{a} \in \prod_{\hbar \in (0,1]} \mathfrak{A}_\hbar$ such that there is an element $a \in \mathfrak{A}$ with

---

$$\tilde{a}(\hbar) = \phi_\hbar(a) \tag{4.2}$$

for each $\hbar \in (0, 1]$. Then we define evaluation maps $\tilde{\phi}_\hbar \colon \tilde{\mathfrak{A}} \to \mathfrak{A}_\hbar$ by

$$\tilde{\phi}_\hbar(\tilde{a}) := \tilde{a}(\hbar) \tag{4.3}$$

for each $\hbar \in (0, 1]$ and $\tilde{a} \in \tilde{\mathfrak{A}}$. We call the resulting structure $((\mathfrak{A}_\hbar, \tilde{\phi}_\hbar)_{\hbar \in (0,1]}, \tilde{\mathfrak{A}})$ the *restriction* of the original continuous bundle of C*-algebras $((\mathfrak{A}_\hbar, \phi_\hbar)_{\hbar \in [0,1]}, \mathfrak{A})$ over the closed interval $[0, 1]$ to the new domain given by the half-open interval $(0, 1]$.

This new structure $((\mathfrak{A}_\hbar, \tilde{\phi}_\hbar)_{\hbar \in (0,1]}, \tilde{\mathfrak{A}})$ is no longer a continuous bundle of C*-algebras because in general there will be elements $\tilde{a} \in \tilde{\mathfrak{A}}$ such that the function

$$\hbar \in (0, 1] \mapsto \|\tilde{\phi}_\hbar(\tilde{a})\|_\hbar \tag{4.4}$$

is continuous, but does not vanish as $\hbar \to 0$, that is, does not belong to $C_0((0, 1])$. Instead, this structure has the following properties, analogous to those of a continuous bundle:

(i) for each $\tilde{a} \in \tilde{\mathfrak{A}}$, $\|\tilde{a}\| = \sup_{\hbar \in (0,1]} \|\tilde{\phi}_\hbar(\tilde{a})\|_\hbar$;

(ii) for each $f \in UC_b((0, 1])$ and each $\tilde{a} \in \tilde{\mathfrak{A}}$, there is a section $f\tilde{a} \in \tilde{\mathfrak{A}}$ such that $\tilde{\phi}_\hbar(f\tilde{a}) = f(\hbar)\tilde{\phi}_\hbar(\tilde{a})$ for each $\hbar \in (0, 1]$;

(iii) for each section $\tilde{a} \in \tilde{\mathfrak{A}}$, the map $\hbar \mapsto \|\tilde{a}\|_\hbar$ belongs to $UC_b((0, 1])$.

Steeger and Feintzeig (2021a,b) call such a structure a *uniformly continuous bundle of C*-algebras* because of the appearance of uniformly continuous functions in the definition. This is the structure we assume as given by the quantum theory. For the moment, we justify this starting point only with the idea that all of this data for values $\hbar > 0$ counts as part of the quantum theory. Our discussion in the next section will lead to further justification for starting with the algebra of uniformly continuous sections $\tilde{\mathfrak{A}}$ as part of the quantum theory, and so we postpone further discussion of this point.

Now let us suppose we are given a *uniformly* continuous bundle of C*-algebras $((\mathfrak{A}_\hbar, \phi_\hbar)_{\hbar \in (0,1]}, \mathfrak{A})$ over the base space $(0, 1]$ representing the quantities in a quantum theory for $\hbar > 0$ (we now remove the tildes for notational convenience). We can reformulate our question as: can one reconstruct the C*-algebra $\mathfrak{A}_0$ of classical quantities from the data contained in the given uniformly continuous bundle? Indeed, just as uniformly continuous functions have unique extensions to limiting values, it turns out that uniformly continuous bundles do as well. Steeger and Feintzeig (2021a) provide the following result.

Theorem 6 *Given a uniformly continuous bundle of C*-algebras $((\mathfrak{A}_\hbar, \phi_\hbar)_{\hbar \in (0,1]}, \mathfrak{A})$ over the base space $(0, 1]$, there is a unique (up to *-isomorphism)*

C*-algebra $\mathfrak{A}_0$, *and a *-homomorphism* $\phi_0 : \mathfrak{A} \to \mathfrak{A}_0$ *such that* $((\mathfrak{A}_\hbar, \phi_\hbar)_{\hbar \in [0,1]},$ $\mathfrak{A})$ *is a continuous bundle of C*-algebras over the enlarged base space* $[0, 1]$.

We think of the structure $((\mathfrak{A}_\hbar, \phi_\hbar)_{\hbar \in [0,1]}, \mathfrak{A})$ as the *extension* of the original (restricted) uniformly continuous bundle of C*-algebras from the base space $(0, 1]$ to $[0, 1]$, thus including the algebra $\mathfrak{A}_0$ at $\hbar = 0$. The strategy of the proof, which we sketch in what follows, is to understand the fibers of a bundle of C*-algebras as quotients of the C*-algebra of continuous sections.[87]

Proof sketch of Theorem 6: Consider the subset

$$K_0 := \left\{ a \in \mathfrak{A} \mid \lim_{\hbar \to 0} \| \phi_\hbar(a) \|_\hbar = 0 \right\}. \tag{4.5}$$

of $\mathfrak{A}$. One can show that K_0 is a closed two-sided ideal in $\mathfrak{A}$. This allows us to define the fiber algebra at $\hbar = 0$ as the quotient C*-algebra

$$\mathfrak{A}_0 := \mathfrak{A} / K_0. \tag{4.6}$$

Further, we can use the canonical quotient map as the evaluation map $\phi_0 : \mathfrak{A} \to \mathfrak{A}_0$ at $\hbar = 0$ by defining

$$\phi_0(a) := [a] \tag{4.7}$$

for all $a \in \mathfrak{A}$, where $[a]$ is the equivalence class of a in $\mathfrak{A} / K_0$. One can show that the resulting structure is indeed a continuous bundle of C*-algebras, and it is unique. $\square$

In this case, the C*-algebraic operations in the classical theory at $\hbar = 0$ are explicitly defined from the C*-algebraic operations at $\hbar > 0$ by the definition of the quotient C*-algebra. Hence, this provides a sense in which the C*-algebraic structure of the physical quantities in the classical theory is determined by that in the corresponding quantum theory.

We pause here to consider a possible objection. One might worry that in Theorem 6, removing reference to the classical algebra $\mathfrak{A}_0$ is not enough. Specifically, one might claim that the use of the algebra $\mathfrak{A}$ of continuous sections already prescribes a scaling dependence based on the $\hbar \to 0$ limit that puts in classical structure by hand. If this is correct, then Theorem 6 would not show that the classical algebra is determined by the algebra of the quantum theory – because we added to the quantum theory extra structure by selecting specific quantities and limiting procedures based on knowledge of the higher level theory.

---

[87] See also Lee (1976).

In response, we admit that it is possible for quantum theories to give rise to different higher-level theories based on different quantities and limiting operations, which would indeed be represented by different algebras $\mathfrak{A}$ of continuous sections. However, we maintain that in some cases – and in the $\hbar \to 0$ limit in particular – one can have knowledge of the type of scaling limit that is, in some sense, independent of the content of the higher-level theory. We will argue in the next section that one can interpret the rescalings of $\hbar$ on the way to the classical limit as changes of units within the quantum theory itself, so that the algebra $\mathfrak{A}$ is determined by the structure of the quantum theory along with an interpretation of quantities as taking values with physical units. If this is correct, then one should not worry that the algebra $\mathfrak{A}$ somehow smuggles in information from classical physics.

Next, we treat the layer of structure represented by the classical Poisson bracket. Can this structure be recovered from a quantum theory? Recall that Dirac's condition guarantees that the Poisson bracket comes to agree with $\frac{i}{\hbar}$ times the commutator in the quantum theory, at least approximately in the limit as $\hbar \to 0$. This, in turn, allows one to understand the classical Poisson bracket as *definable* from the commutator in the quantum theory. To show this, we build on the previous construction of classical C*-algebraic structure with the following result.

Theorem 7 *Suppose* $((\mathfrak{A}_\hbar, \phi_\hbar)_{\hbar \in (0,1]}, \mathfrak{A})$ *is a uniformly continuous bundle of C*-algebras over* $(0, 1]$ *and* $((\mathfrak{A}_\hbar, \phi_\hbar)_{\hbar \in [0,1]}, \mathfrak{A})$ *is its extension to* $[0, 1]$ *as guaranteed by Theorem 6. Then for* $a, b, c \in \mathfrak{A}$ *satisfying*

$$\phi_\hbar(c) = \frac{i}{\hbar}[\phi_\hbar(a), \phi_\hbar(b)], \tag{4.8}$$

for all $\hbar > 0$, *the operation*

$$\{\phi_0(a), \phi_0(b)\} := \phi_0(c) \tag{4.9}$$

defines a Poisson bracket on a subset of $\mathfrak{A}_0$. *Moreover, if the extended bundle of C*-algebras is one generated by a strict quantization of a Poisson algebra* $\mathcal{P}$ *as in Prop. 1, then the bracket defined by Eq. (4.9) agrees with the Poisson bracket on* $\mathcal{P}$.

This establishes a sense in which the Poisson bracket of the classical theory at $\hbar = 0$ is completely determined by the commutator in the quantum theory at $\hbar > 0$ through the $\hbar \to 0$ limit.

Finally, we treat dynamical structure. For this, we understand Hamiltonian dynamics for some smooth, real-valued Hamiltonian function $h : M \to \mathbb{R}$ on a phase space M to give rise to a one-parameter family of automorphisms

$(\tau_{t;0})_{t\in\mathbb{R}}$ of the classical C*-algebra $\mathfrak{A}_0$ of functions on M satisfying the Hamiltonian equations of motion

$$\frac{d}{dt}\tau_{t;0}(A) = \{h, A\} \qquad (4.10)$$

for each quantity A in the Poisson subalgebra $\mathcal{P}$ of $\mathfrak{A}_0$. Similarly, we understand the Heisenberg picture of the quantum dynamics for $\hbar > 0$ to give rise to a one-parameter family $(\tau_{t;\hbar})_{t\in\mathbb{R}}$ of automorphisms of the quantum C*-algebra $\mathfrak{A}_\hbar$ of operators. When $\mathfrak{A}_\hbar$ is represented on a Hilbert space $\mathcal{H}$ via a suitable *-homomorphism π, this dynamics will satisfy the Heisenberg equations of motion

$$\pi(\tau_{t;\hbar}(B)) = e^{-iHt}\pi(B)e^{iHt} \qquad (4.11)$$

for a (typically unbounded) Hamiltonian operator H. The following result builds on the previous two theorems to show a sense in which the quantum dynamics determines the classical dynamics whenever the quantum dynamics vary continuously with $\hbar$.

Theorem 8 *Suppose $((\mathfrak{A}_\hbar, \phi_\hbar)_{\hbar\in(0,1]}, \mathfrak{A})$ is a uniformly continuous bundle of C*-algebras over $(0,1]$ and $((\mathfrak{A}_\hbar, \phi_\hbar)_{\hbar\in[0,1]}, \mathfrak{A})$ is its extension to $[0,1]$ as guaranteed by Theorem 6. Suppose, moreover, for each $\hbar > 0$, there is a one-parameter family $(\tau_{t;\hbar})_{t\in\mathbb{R}}$ of automorphisms of $\mathfrak{A}_\hbar$. If these automorphisms lift to a one-parameter family of automorphisms $(\tau_t)_{t\in\mathbb{R}}$ of $\mathfrak{A}$ satisfying*

$$\phi_\hbar \circ \tau_t(a) = \tau_{t;\hbar} \circ \phi_\hbar(a) \qquad (4.12)$$

for all $a \in \mathfrak{A}$, then there is a unique one-parameter family of automorphisms $(\tau_{t;0})_{t\in\mathbb{R}}$ of $\mathfrak{A}_0$ satisfying

$$\phi_\hbar \circ \tau_t(a) = \tau_{t;0} \circ \phi_\hbar(a) \qquad (4.13)$$

for all $a \in \mathfrak{A}$.

Moreover, suppose the extended bundle of C-algebras is one generated by Weyl quantization on $C_c^\infty(\mathbb{R}^{2n})$ (see Ex. 16). Suppose that $(\tau_{t;\hbar})_{t\in\mathbb{R}}$ satisfies the Heisenberg dynamics in Eq. (4.11) for each $\hbar > 0$ with a Hamiltonian operator $H = \pi_S(Q_\hbar(h))$ determined by a classical Hamiltonian $h \in C_c^\infty(\mathbb{R}^{2n})$. Then it follows that $(\tau_{t;0})_{t\in\mathbb{R}}$ satisfies the Hamiltonian dynamics in Eq. (4.10).*

This establishes a sense in which classical dynamics at $\hbar = 0$ are determined by quantum dynamics for $\hbar > 0$ in the $\hbar \to 0$ limit. We remark that although the statement of the preceding theorem is somewhat limited in its application to only smooth and bounded Hamiltonians, the same result can also be extended to many other classical Hamiltonians – for example, Hamiltonians that are unbounded polynomials quadratic in the position and momentum variables.[88] So the result covers a range of physically significant dynamics.

The foregoing suggests that the classical $\hbar \to 0$ limit is not "singular" – at least in the sense that a quantum theory on its own *does* contain enough information to determine the structure of a classical theory. That is, Theorem 6 shows a sense in which the C*-algebraic structure of the physical quantities in a quantum theory determines the C*-algebraic structure of the physical quantities in the corresponding classical theory through the $\hbar \to 0$ limit. Theorem 7 shows a sense in which the commutator Lie bracket in a quantum theory determines the Poisson bracket in the corresponding classical theory through the $\hbar \to 0$ limit. And finally, Theorem 8 shows a sense in which the dynamics of a quantum theory determines the dynamics of the corresponding classical theory through the $\hbar \to 0$ limit. One needs no information for these results beyond that defined on the uniformly continuous bundle of C*-algebras for $\hbar > 0$ representing the quantum theory. Thus, one does not require extra information from the classical theory to make the $\hbar \to 0$ limit well-defined.

4.3 Approximation

The purpose of this section is to clarify the sense in which the determination of classical structure from quantum structure can be understood to explain the success of classical physics on the basis of quantum physics. We will do so by making precise an interpretation in which the theoretical structure of classical physics provides an *approximation* to the theoretical structure of quantum physics. Our discussion will yield two further consequences: first, it will provide a response to a well-known objection to the coherence of varying values of $\hbar$, and second, it will provide a justification for taking an entire uniformly continuous bundle of C*-algebras over $(0, 1]$ to be determined solely by the information in a quantum theory for $\hbar > 0$.

We will motivate our concern in this section by considering the aforementioned objection to the coherence of the $\hbar \to 0$ limit. This objection has been voiced by a number of authors:

> Indeed, it makes no physical sense to permit physical constants to vary ($\hbar \to 0, c \to \infty$). (Nickles, 1973, p. 201)

> Of course, from a physical point of view putting $h = 0$ is impossible: h is an empirically fixed constant unequal to zero[.] (Radder, 1991, p. 209, fn. 8)

---

[88] See Landsman (1998a, p. 148) for generalizations of Theorem 8.

One may make the predictable criticism that Batterman's analysis relies on the limit $\hbar \to 0$ even though $\hbar$ is constant for all real systems, and that the relevance of the analysis for real systems is obscured by this fact. (Rosaler, 2015a, p. 331)

The idea seems to be that a different value of $\hbar$ would correspond to a different possible world with different laws of physics. No matter what we could show about such a world, it could not explain anything about *our* world.

There is a standard reply that the limit $\hbar \to 0$ is elliptical for considering small values of the ratio of $\hbar$ to some other properties of a physical system. For example, Batterman writes:

> One might reasonably wonder what it could mean to let a constant change its value. The way to understand "$\hbar \to 0$" is that it is the limit in which $\hbar$ is small relative to some quantity having the same dimension – namely, the classical action. (Batterman, 2002, p. 99, fn. 1)

Indeed, Landsman (2017, p. 247) takes a similar stance. He suggests that the particular dimensionless ratio we consider may depend on context, but that, for example, in recovering the classical limit of Planck's blackbody radiation law, we should read the $\hbar \to 0$ limit as the limit $\frac{h\nu}{kT} \to 0$, which can be achieved through either letting $T \to \infty$ or $\nu \to 0$. These correspond to different families of physically possible states in *our* world. Or, for another example, one might consider the classical limit of nonrelativistic quantum particle dynamics defined by a bounded potential $V(x)$ with a typical energy scale $\epsilon := \sup_x |V(x)|$ and length scale $\ell := \epsilon / \sup_x |\nabla V(x)|$ (supposing the derivatives of V are bounded). In this case, he suggests we should read the $\hbar \to 0$ limit as the limit $\frac{\hbar}{\ell\sqrt{2m\epsilon}} \to 0$, which can be achieved through letting $\ell \to \infty$ or $\epsilon \to \infty$. These correspond to different families of physically possible dynamics determined by physically possible potentials, again in *our* world.

However, treating the classical limit in terms of small dimensionless ratios of $\hbar$, while a viable strategy, has a number of drawbacks. First, as Rosaler (2015a, p. 331) argues, since mathematical results on the $\hbar \to 0$ limit rarely explicitly treat a dimensionless ratio of $\hbar$ to some quantity with the units of classical action, we lose track of the very conditions under which the results are supposed to be applicable to real physical systems. Second, while the mathematical tools of strict quantization *can* be used to treat limits of states and dynamics that continuously vary between the fiber C*-algebras, there is an important prior notion of convergence as the algebraic structure of the physical quantities itself varies continuously with $\hbar$. The sense in which we understand kinematic relations like the canonical commutation relations to vary with $\hbar$ cannot be captured by an interpretation in which varying $\hbar$ corresponds to varying quantities dependent on particular states or dynamics. Third, and building on the last point, while varying states and dynamics with $\hbar$ might lead to an explanation of why the empirical predictions of a given classical state or dynamics are approximately correct and so might yield what we have called an empirical reduction, it is not clear how this might lead to what we have called a theoretical reduction, or an explanation of the theoretical structure of a classical theory. Each of these issues, however, can be addressed with a different interpretation of the $\hbar \to 0$ that does not treat $\hbar$ as elliptical for some dimensionless ratio.

The remainder of this section will outline an interpretation of the $\hbar \to 0$ limit that provides a precise sense in which the empirical predictions of a classical theory can be approximated *uniformly* for all states from those of a quantum theory. This leads to an explanation of the theoretical structure of classical physics – including the kinematical structure – and is applicable to *all* systems in our world, regardless of whether they are in a state with a large typical classical action.

The central idea of the interpretation of the classical limit offered in the remainder of this section is that different values of $\hbar$ are interpreted as numerical values of Planck's constant in different systems of units.[89] Each value $\hbar \in (0, 1]$ then corresponds to a system of units one can use in *our* world. For example, Planck's constant in our world can take either the numerical value $6.626\ldots \times 10^{-34}$ $m^2 \cdot kg/s$, or – changing units from m to km – the value $6.626\ldots \times 10^{-40}$ $km^2 \cdot kg/s$ (since $1 \ m^2 = 1 \times 10^{-6} \ km^2$). Conversely, if we restrict ourselves to only changes of distance units, then one can determine how physical quantities with distance units change by the change in Planck's constant: if Planck's constant changes by a factor of 10^{-6}, then all distances must change by a factor of 10^{-3}. More generally, since Planck's constant has units [$distance^2 \cdot mass/time$], one can determine the changes in physical quantities with units of distance, mass, and time from changes in Planck's constant, as long as the form of the unit change is stipulated at the outset (e.g., we agree to change only time units, or we agree to change distance and time units at the same rate, etc.).

Even with this interpretation in mind, it may still seem mysterious how such unit changes could play a role in the classical limit and how they might connect to strict quantization. We will clarify those connections in what follows.

In order to connect such unit changes to strict quantization, we will consider a concrete example in the quantization of the Weyl algebra. Let us consider how

---

[89] For more discussion, see Feintzeig (2020).

a change of units affects the Weyl unitaries for an n-particle quantum system in the Schrödinger representation. We will stick with changes of distance units, although the same analysis can be applied to other unit changes. Recall that the Weyl unitaries for a system with phase space $\mathbb{R}^{2n}$ take the form

$$\pi_S(W_\hbar(a, b)) = e^{i \sum_{j=1}^n a_j \sqrt{\hbar} \cdot P_j + b_j \sqrt{\hbar} \cdot Q_j}. \tag{4.14}$$

acting on $L^2(\mathbb{R}^n)$. But since each Q_j has units of [*distance*], and each P_j has units of [*mass × distance/time*], we know how a change of distance units will change these operators, and hence the Weyl unitaries. For example, changing distance units (e.g., from m to km) will change the values (spectrum) of Q_j and P_j by a factor of α, leading to the transformations

$$\begin{aligned} Q_j &\mapsto \alpha \cdot Q_j \\ P_j &\mapsto \alpha \cdot P_j. \end{aligned} \tag{4.15}$$

But since $\hbar$ has units [*distance^2 × mass/time*], this also leads to the transformation

$$\hbar \mapsto \alpha^2 \cdot \hbar. \tag{4.16}$$

If we denote the transformed value of Planck's constant in the new system of units by $\hbar' := \alpha^2 \cdot \hbar$, then we have $\alpha = \sqrt{\frac{\hbar'}{\hbar}}$ as the scaling factor for a distance unit change.

To make the analysis relevant to *our* world, we recall that different values of $\hbar$ in the Weyl algebra $\mathcal{W}(\mathbb{R}^{2n}, \hbar\sigma)$ correspond to different canonical commutation relations (in Eq. (2.37)) encoding substantive kinematic structure. Let us fix the value of Planck's constant for the kinematic structure of our world in so-called "natural units" with $\hbar_0 = 1$. Then, working in the Weyl algebra $\mathcal{W}(\mathbb{R}^{2n}, \sigma)$ for our world, consider the trajectories obtained by looking at the same physical quantities as the Weyl unitaries in different systems of units. If we change units from $\hbar_0 = 1$ to $\hbar$, then we have the scaling factor $\alpha = \sqrt{\hbar}$. The unit change then corresponds to the transformation

$$\begin{aligned} Q_j &\mapsto \sqrt{\hbar} \cdot Q_j \\ P_j &\mapsto \sqrt{\hbar} \cdot P_j, \end{aligned} \tag{4.17}$$

which in turn affects the transformation of the Weyl unitaries in the Schrödinger representation via the following transformation:

$$\pi_S(W_1(a, b)) = e^{i \sum_{j=1}^n a_j P_j + b_j Q_j} \mapsto e^{i \sum_{j=1}^n a_j \sqrt{\hbar} \cdot P_j + b_j \sqrt{\hbar} \cdot Q_j} = \pi_S(W_1(\sqrt{\hbar} \cdot a, \sqrt{\hbar} \cdot b)). \tag{4.18}$$

Under these transformations representing distance unit changes, we can understand all of the operators in $\mathcal{W}(\mathbb{R}^{2n}, \sigma)$ along the unit-change trajectories

$$\hbar \mapsto W_1(\sqrt{\hbar} \cdot a, \sqrt{\hbar} \cdot b) \tag{4.19}$$

as representing the same physical quantities in different systems of units. Finally, one can notice that the map $\beta_\hbar : \mathcal{W}(\mathbb{R}^{2n}, \sigma) \to \mathcal{W}(\mathbb{R}^{2n}, \hbar\sigma)$ defined as the continuous linear extension of the unit-change map

$$\beta_\hbar\big(W_1(\sqrt{\hbar} \cdot a, \sqrt{\hbar} \cdot b)\big) := W_\hbar(a, b) \tag{4.20}$$

is a *-isomorphism.[90] This last fact allows us to think of the curves traced out in the space $\prod_{\hbar \in (0,1]} \mathcal{W}(\mathbb{R}^{2n}, \hbar\sigma)$ by the maps

$$\hbar \in (0, 1] \mapsto W_\hbar(a, b) \tag{4.22}$$

as connecting points that represent the same physical quantity (under the standard of comparison given by the family of *-isomorphisms $\beta_\hbar$ in Eq. (4.20)). But these are precisely the curves traced out by the Weyl quantization maps and so provide precisely the structure defined by a strict quantization.

We are now in a position to justify the assumption in the previous section that a quantum theory on its own determines a bundle of C*-algebras over $(0, 1]$ for values of $\hbar > 0$. Notice that our analysis thus far has taken place entirely within a quantum theory. Thus, a quantum theory with enough structure to represent unit changes determines the curves in Eq. (4.22). Those curves generate a C*-algebra $\mathfrak{A} \subseteq \prod_{\hbar \in (0,1]} \mathcal{W}(\mathbb{R}^{2n}, \hbar\sigma)$. This is precisely the algebra of sections of the uniformly continuous bundle of C*-algebras over $(0, 1]$ that is used as the starting point of Theorems 6–8 in the previous section. Since our analysis here of changes of units has taken place entirely in a quantum theory, we have shown that *if a quantum theory of n particles comes with the structure to identify the same physical quantity in different systems of units*, then that quantum theory determines the uniformly continuous bundle of C*-algebras $((\mathcal{W}(\mathbb{R}^{2n}, \hbar\sigma), \phi_\hbar)_{\hbar \in (0,1]}, \mathfrak{A})$ for values $\hbar > 0$, which in turn corresponds to the strict quantization of the Weyl algebra.

With this understanding of different values of $\hbar$ corresponding to different systems of units, we can finally state the precise sense in which the strict quantization of the Weyl algebra establishes that the theoretical structure of classical

---

[90] Another way to see this concretely is that Eq. (4.18) implies

$$\pi_S(W_1(\sqrt{\hbar} \cdot a, \sqrt{\hbar} \cdot b)) = \pi_S(W_\hbar(a, b)), \tag{4.21}$$

where the same symbol π_S is used to denote the Schrödinger representation of $\mathcal{W}(\mathbb{R}^{2n}, \sigma)$ on the left and $\mathcal{W}(\mathbb{R}^{2n}, \hbar\sigma)$ on the right. Since the Weyl algebra is simple, in both cases π_S is faithful, which implies that $\beta_\hbar$ is a *-isomorphism.

physics approximates that of quantum physics. We consider the content of each of Dirac's, von Neumann's, and Rieffel's conditions on a strict quantization. Those limiting conditions state that for all classical physical quantities $A, B \in \mathcal{P}$ and for any numerical error bound $\epsilon > 0$, there is some system of units in which Planck's constant takes a small enough value $\hbar > 0$ such that the algebraic relations of the corresponding physical quantities $Q_\hbar^W(A), Q_\hbar^W(B) \in \mathcal{W}(\mathbb{R}^{2n}, \hbar\sigma)$ are close to those of the classical quantities in the sense that:

$$\left\| \frac{i}{\hbar}[Q_\hbar^W(A), Q_\hbar^W(B)] - Q_\hbar^W(\{A, B\}) \right\|_\hbar < \epsilon$$
$$\left\| Q_\hbar^W(A) Q_\hbar^W(B) - Q_\hbar^W(AB) \right\|_\hbar < \epsilon \tag{4.23}$$
$$\left| \|Q_\hbar^W(A)\|_\hbar - \|A\|_0 \right| < \epsilon.$$

The fact that the first two inequalities are norm differences implies that ϵ is a uniform error bound such that *for all states* ω on $\mathcal{W}(\mathbb{R}^{2n}, \hbar\sigma)$,

$$\left| \omega\left(\frac{i}{\hbar}[Q_\hbar^W(A), Q_\hbar^W(B)] \right) - \omega(Q_\hbar^W(\{A, B\})) \right| < \epsilon$$
$$\left| \omega(Q_\hbar^W(A) Q_\hbar^W(B)) - \omega(Q_\hbar^W(AB)) \right| < \epsilon. \tag{4.24}$$

Further, these inequalities will hold in any system of units in which Planck's constant takes a numerical value smaller than $\hbar$. The value of $\hbar$ and the number ϵ together determine the scale at which the classical theory approximates the quantum theory. Namely, the value of $\hbar$ corresponds to the system of units in which one should interpret the number ϵ, which then represents the unitful, and physical, degree of approximation between the corresponding quantities and expectation values. Of course, for a choice of $\epsilon > 0$, the value of $\hbar$ for which the inequalities (4.23) and (4.24) hold depends on the particular quantities A and B that one is interested in approximating. So while the approximation is uniform over the expectation values of all *states*, it is not generally uniform over all choices of *quantities*.

We mention also that although we have focused on the approximation of classical kinematical structure, the results of Theorems 7 and 8 imply that similar uniform or state-independent approximations hold for suitable dynamics. To see this, suppose that we are given a family of automorphisms $(\tau_{t;\hbar})_{t\in\mathbb{R}}$ for each algebra $\mathcal{W}(\mathbb{R}^{2n}, \hbar\sigma)$ with $\hbar \in [0, 1]$. If there exists a family of automorphisms $(\tau_t)_{t\in\mathbb{R}}$ of the algebra of continuous sections lifting the dynamics on the fibers in the sense of satisfying Eq. (4.12), then for any $A \in \mathcal{P}$, the map $a := [\hbar \mapsto \tau_t(Q_\hbar(A))]$ is a continuous section with

$$\phi_0(a) = \tau_{t;0}(A) \tag{4.25}$$

given by the classical dynamics and

$$\phi_\hbar(a) = \tau_{t;\hbar}(Q_\hbar^W(A)) \tag{4.26}$$

given by the quantum dynamics. Now, since $[\hbar \mapsto Q_\hbar^W(\tau_{t;0}(A))]$ and $[\hbar \mapsto \tau_{t;\hbar}(Q_\hbar^W(A))]$ are both continuous sections with the same value at $\hbar = 0$, it follows that

$$\lim_{\hbar \to 0} \|Q_\hbar^W(\tau_{t;0}(A)) - \tau_{t;\hbar}(Q_\hbar^W(A))\|_\hbar = 0. \tag{4.27}$$

Thus, it follows that for every numerical error bound $\epsilon > 0$, there is some system of units in which Planck's constant takes a small enough value $\hbar > 0$ such that

$$\|Q_\hbar^W(\tau_{t;0}(A)) - \tau_{t;\hbar}(Q_\hbar^W(A))\|_\hbar < \epsilon. \tag{4.28}$$

In turn this implies that ϵ is a uniform error bound such that for every state ω on $\mathfrak{A}_\hbar$,

$$|\omega(Q_\hbar^W(\tau_{t;0}(A))) - \omega(\tau_{t;\hbar}(Q_\hbar^W(A)))| < \epsilon. \tag{4.29}$$

As before, these inequalities will also hold in any system of units in which Planck's constant takes a numerical value smaller than $\hbar$. Since the first term is the expectation value of the classical evolution of A and the second term is the expectation value of the quantum evolution of A, this provides a uniform approximation that is a state-independent generalization of the approximation from Rosaler in Eq. (4.1). (In fact, this argument is general enough to apply to arbitrary strict quantizations, not just the strict quantization of the Weyl algebra.)

Finally, one might wish to consider how the interpretation of variable $\hbar$ that we provide here relates to the aforementioned interpretations by, for example, Batterman (2002) and Landsman (2017) that varying $\hbar$ is elliptical for varying some other dimensionless ratio of quantities. While we have advocated for interpreting varying $\hbar$ in terms of different systems of units for selecting the continuous sections, and hence for explaining the algebraic structure of classical mechanics, this still leaves open the explanation of other classical structures. We believe that varying dimensionless ratios can be used in explanations of particular classical dynamics or particular classical states. For example, one rarely takes the classical limit of a quantum state by considering the *same* quantum state for each value of $\hbar$. Instead, the quantum state varies with $\hbar$ (depending on the particular physical context), and this variation may be understood in terms of other state-dependent quantities like temperature T as it figures into the dimensionless ratio $\frac{h\nu}{kT}$ that Landsman (2017) considers. Similarly, one rarely takes the classical limit of a quantum dynamics by considering the *same* quantum dynamics for each value of $\hbar$. Instead, the quantum dynamics varies with $\hbar$ (again depending on the particular physical context), and this variation may be understood in terms of dynamical parameters in the potential like the maximum height ϵ as it figures into the dimensionless ratio $\frac{\hbar}{\ell\sqrt{2m\epsilon}}$ that Landsman (2017) considers.

It is difficult to see how varying values of T and ϵ would have any effect on the kinematic relations codified in the algebraic structure, and so the interpretation of $\hbar$ in terms of changes of units helps us understand the explanation of classical algebraic structure via continuous bundles. However, we see this explanation as perfectly compatible with the idea that varying, for example, T and ϵ selects relevant families of states and dynamics that can be used to explain classical states and dynamics, once the algebraic structure has been determined. In other words, we do not see a conflict between these different ways of interpreting varying values of $\hbar$; rather, we see the different interpretations playing different roles, and working together to achieve compatible explanatory ends.

Thus, we conclude that (i) varying values of $\hbar$ can be coherent and explanatory under the interpretation of unit changes outlined here, (ii) that when a quantum theory is understood to incorporate such unit changes, it comes with the structure of a bundle of C*-algebras over values $\hbar > 0$, and (iii) this interpretation gives rise to a precise sense in which the theoretical structure of classical physics approximates that of quantum physics.

5 Interpretation

The next philosophical topic we shall consider is the interpretation of quantum theories. We will suggest that the classical $\hbar \to 0$ limit might provide tools to aid the interpretive process. To do so, we explore two uses of the classical limit.

The first topic enters in the interpretational debates concerning scientific realism. Various authors, going back at least to Worrall (1989), have advocated for a position called *structural realism*. And others (French and Ladyman, 2003) have even advocated for applying a similar perspective specifically to quantum theories. One central line of argument for the structural realist position relies on (often vague) claims about diachronic structural continuity – in our case, the claim that some structure is preserved in the transition from classical to quantum physics. We will show that the $\hbar \to 0$ limit provides some tools for assessing those claims.

Our second topic is the role of analogies in interpreting quantum theories. Hesse (1970) famously advocated for the role of analogies in interpreting scientific theories generally. We will take up this suggestion in two case studies. First, we analyze the interpretation of a continuity condition callled *regularity* that is often taken to restrict the physically possible quantum states. Since the regularity condition has been challenged as being unjustified, we will seek to analyze its significance and justification by looking at the analogous condition in classical theories. Second, we analyze the purported interpretational problems raised by inequivalent particle notions in quantum field theory. Our strategy is to look for analogs in a corresponding classical field theory of the mathematical structures that underlie the inequivalent particle notions in the quantum theory. Since (as we hope it is uncontroversial) classical field theories are at least more familiar and better understood than quantum field theories, we hope to gain some insight from the classical theory that might inform our interpretation of the quantum theory.

Throughout this section, we will avoid discussion of some of the traditional interpretive questions in quantum theories concerning measurement and locality. In doing so, we suggest that there is much fertile philosophical ground beyond those contested issues.[91]

5.1 Structural Continuity

Worrall (1989) proposes a version of scientific realism that highlights the *mathematical structure* of scientific theories as the content we should expect to be preserved across theory change. Worrall's position aims to make sense of the success of scientific inquiry, which, for example, Boyd (1973, 1989) takes to be evidence for scientific realism. However, Worrall also claims his position makes sense of the deep theoretical shifts that have occurred throughout the history of science, which, for example, Laudan (1981) and Stanford (2006) have taken as evidence against scientific realism. The central contention of Worrall's structural realism is that even when there are deep theoretical shifts in scientific thinking, some structure is still preserved from the old theories to the new ones. Worrall's own example is the preservation of Fresnel's equations governing the reflection and refraction of light based on a conception of light as waves in the ether. Fresnel's equations retain their form in Maxwell's theory of light as electromagnetic waves. Although, Worrall claims, Maxwell's theory rid itself of the theoretical posit of the ether,[92] it retained important structure from Fresnel's theory that should form the basis of our physical interpretation and beliefs.

---

[91] This is also a central theme of Ruetsche (2011), which we simply echo here.

[92] While Worrall attributes this to Maxwell, the history is considerably more complicated and perhaps Worrall's claims are better understood by replacing Maxwell's with Einstein's electrodynamics. See, for example, Stein (1981, 1987) for more detail.

Proponents of structural realism in general believe that this position applies to a broad class of scientific theories whenever we observe diachronic continuity in theoretical structure. Although Worrall (1989, p. 123) himself expressed some skepticism concerning the applicability of this view to quantum physics,[93] others (Ladyman, 1998; French and Ladyman, 2003) have advocated for applying the view to the quantum realm in some form.[94] Perhaps surprisingly, many of the motivations cited by those authors for taking a structural realist position toward quantum theory completely leave behind the diachronic structural continuity that Worrall began with. One might wonder whether the case of quantum theory can be understood to fit with Worrall's original argument.

We take this to suggest an interesting question: *to what extent is structure preserved in the transition from classical to quantum physics?* This is the question we take up in the remainder of this section. We will not advocate for structural realism here, nor do we think one must find it a plausible position to be interested in this question. We take the foregoing only to situate our investigation of preserved structure. Our strategy is to leave the understanding of the significance of preserved structure across theoretical change for after one has assessed the precise form of any structural continuities.

Previous work (Thébault, 2016; Yaghmaie, 2021) has suggested that quantization might provide tools for understanding the structure preserved from classical to quantum physics. Indeed, one might also take strict quantizations and their associated continuous bundles of C*-algebras to already exhibit a relevant continuity of structure between classical and quantum. Here, we will confirm these suggestions in some select cases and demonstrate a method for making structural comparisons across theory change more precise.

The method we employ to analyze theoretical structure is to understand structure as *that which is invariant under structure-preserving maps between mathematical models.* This allows us to compare the structure-preserving maps between models of quantum physics with the structure-preserving maps between models of classical physics. Whereas the models themselves – for example, represented by commutative and noncommutative C*-algebras, respectively – are importantly distinct, we will show that the structure-preserving maps can be translated from classical to quantum theories, and vice

---

[93] The discussion in Worrall (1989, p. 123) seems to make the assumption that any classical structure reappearing in quantum theories would need to take the form of a hidden variable theory. This leaves open the possibility that we investigate here, that the standard mathematical structures of classical and quantum physics might be continuous with one another.

[94] For a small sampling of contemporary discussions of structural realism, see, for example, Bueno (1999), Psillos (2001), Cao (2003), Stanford (2003, chapter 7), Brading and Landry (2006), French and Saatsi (2006), Frigg and Votsis (2011).

versa. This provides a precise sense in which classical and quantum theories have shared structure.

Why should one care about making structural continuity precise in this way? One virtue of our approach is that it highlights two different aspects of these claims about structural comparisons corresponding to different directions – from classical to quantum, or from quantum to classical. Ultimately, we will characterize the strongest kind of structural preservation, which holds when the translations of structure-preserving maps in each direction are inverse to each other. By using these tools, our approach to structural continuity across theory change makes substantial connections with other work on structural comparisons in philosophy of science, mostly done in the context of analyzing theoretical equivalence.[95]

We will focus again on just one example: the quantization of the Weyl algebra. We will specify the structure-preserving maps in the classical and quantum theories, and then we will establish a correspondence between them.

First, we specify the structure-preserving maps of the classical theory. Suppose (V_1, σ_1) and (V_2, σ_2) are symplectic topological vector spaces representing the test function spaces of two classical models. Recall that the classical algebras of observables for these theories are $AP(V_1')$ and $AP(V_2')$ with Poisson subalgebras $\Delta(V_1)$ and $\Delta(V_2)$, respectively. We take structure-preserving maps between the classical models to be *-homomorphisms $\varphi_0: AP(V_1') \to AP(V_2')$ satisfying:

(i) $\varphi_0[\Delta(V_1)] \subseteq \Delta(V_2)$; and
(ii) $\varphi_0(\{W_0(f), W_0(g)\}) = \{\varphi_0(W_0(f)), \varphi_0(W_0(g))\}$ for any $f, g \in V_1$.

These conditions ensure that φ_0 preserves the Poisson structure in addition to the C*-algebraic structure of the physical quantities. Indeed, we are justified in understanding φ_0 as preserving classical structure in part because, as we will see later on, such a map corresponds to a transformation between the phase spaces V_2' and V_1' (or a transformation between the test function spaces V_1 and V_2).

Next, we must specify the structure-preserving maps in the corresponding quantum theories. Recall that Weyl quantization provides maps $Q_\hbar^W: AP(V_1') \to \mathcal{W}(V_1, \hbar\sigma_1)$ and $Q_\hbar^W: AP(V_2') \to \mathcal{W}(V_2, \hbar\sigma_2)$. We take structure-preserving maps between the quantum models to be *-homomorphisms $\varphi_\hbar: \mathcal{W}(V_1, \hbar\sigma_1) \to \mathcal{W}(V_2, \hbar\sigma_2)$ satisfying:

$$\varphi_\hbar[Q_\hbar^W(\Delta(V_1))] \subseteq Q_\hbar^W(\Delta(V_2)). \tag{5.1}$$

[95] For some relevant work on structural comparison for determining theoretical equivalence, see Halvorson (2016), Weatherall (2021, 2019a,b), Barrett (2020), Rosenstock et al. (2015), Rosenstock and Weatherall (2016), Hudetz (2019), and Dewar (2017, 2019, 2022).

This ensures that $\varphi_\hbar$ preserves the "smooth" structure of the model – encoded in the dense subalgebras $Q_\hbar^W(\Delta(V_1)) \subseteq \mathcal{W}(V_1, \hbar\sigma)$ and $Q_\hbar^W(\Delta(V_2)) \subseteq \mathcal{W}(V_2, \hbar\sigma_2)$ – in addition to the C*-algebraic structure of the physical quantities.

We will need to further restrict attention to certain structure-preserving maps in the quantum theory. Our last restriction will ensure that we focus on morphisms of models of the quantum theory that preserve structure in the same way for all values of $\hbar > 0$. Suppose $\varphi_1 : \mathcal{W}(V_1, \sigma_1) \rightarrow \mathcal{W}(V_2, \sigma_2)$ is a structure-preserving map, that is, a *-homomorphism satisfying Eq. (5.1). Then for any $f \in V_1$, we have

$$\varphi_1(W_1(f)) = \sum_{k=1}^{n} c_k W_1(f_k) \tag{5.2}$$

for some finite collection of constants $c_1, \dots, c_n \in \mathbb{C}$ and test functions $f_1, \dots, f_n \in V_2$. For $\hbar > 0$, define the map $\varphi_\hbar : \mathcal{W}(V_1, \hbar\sigma_1) \rightarrow \mathcal{W}(V_2, \hbar\sigma_2)$ by

$$\varphi_\hbar(W_\hbar(f)) := \sum_{k=1}^{n} c_k W_\hbar(f_k), \tag{5.3}$$

and similarly for each $f \in V_1$. (Recall that the Weyl unitaries are a basis for $Q_\hbar^W(\Delta(E_1))$ so this decomposition is unique and $\varphi_\hbar$ is well-defined.) Since $\varphi_\hbar$ is clearly linear, it has a unique continuous linear extension to its entire domain. We say that φ_1 is a *scaling* morphism if for each $\hbar > 0$, the map $\varphi_\hbar$ is a *-homomorphism. In other words, scaling morphisms of quantum theories are those that can be translated to all values of $\hbar > 0$ and still yield a morphism of the quantum theory.[96]

We can now show that the structure-preserving maps between the classical models are in bijective correspondence with the structure-preserving scaling maps between the quantum models. This correspondence is provided through the classical limit in two steps. First, we take these translated structure-preserving maps $\varphi_\hbar$ for values $\hbar > 0$ and lift them to a *-homomorphism between the C*-algebras of continuous sections of the corresponding bundles of C*-algebras over $(0, 1]$ for V_1 and V_2, which we denote $\mathfrak{A}_1$ and $\mathfrak{A}_2$, respectively. Here, we understand $\mathfrak{A}_1$ and $\mathfrak{A}_2$ to be the C*-algebras generated by the sections

$$\begin{bmatrix} \hbar \in (0, 1] \mapsto W_\hbar(f_1) \in \mathcal{W}(V_1, \hbar\sigma_1) \end{bmatrix}$$
$$\begin{bmatrix} \hbar \in (0, 1] \mapsto W_\hbar(f_2) \in \mathcal{W}(V_2, \hbar\sigma_2) \end{bmatrix} \tag{5.4}$$

---

[96] We leave it open whether there are any structure-preserving maps between quantum models that fail to satisfy the scaling condition, or whether there are other natural constraints that imply the scaling condition.

for fixed $f_1 \in V_1$ and $f_2 \in V_2$. To lift $\varphi_\hbar$ to a map between sections in $\mathfrak{A}_1$ and $\mathfrak{A}_2$, we define $\varphi \colon \mathfrak{A}_1 \to \mathfrak{A}_2$ by

$$\phi_\hbar(\varphi(a)) := \varphi_\hbar(\phi_\hbar(a)) \tag{5.5}$$

for any $a \in \mathfrak{A}_1$ and $\hbar \in (0, 1]$. In other words, φ is the transformation on sections generated by the transformations $\varphi_\hbar$ on the values the sections take in the fiber $\mathcal{W}(V_1, \hbar\sigma_1)$. The form of the Weyl quantization map, which generates the algebras of continuous sections $\mathfrak{A}_1$ and $\mathfrak{A}_2$, guarantees that if $a \in \mathfrak{A}_1$ is a continuous section, then $\varphi(a) \in \mathfrak{A}_2$ is a continuous section, too. Thus, given only a structure-preserving map φ_1 for a single fixed value $\hbar = 1$ as our starting point, we now have a *-homomorphism φ between algebras of continuous sections of a bundle representing the classical limit.

Second, we can take the $\hbar \to 0$ limit of a morphism between algebras of continuous sections by employing the construction of the classical limit in Theorem 6. To do so, recall that the fiber algebras at $\hbar = 0$ are isomorphic to quotient algebras. Specifically, if we define

$$\overset{1}{K_0} := \{a \in \mathfrak{A}_1 \mid \lim_{\hbar \to 0} \|\phi_\hbar(a)\|_\hbar = 0\}$$
$$\overset{2}{K_0} := \{a \in \mathfrak{A}_2 \mid \lim_{\hbar \to 0} \|\phi_\hbar(a)\|_\hbar = 0\}, \tag{5.6}$$

then $AP(V_1')$ and $AP(V_2')$ are *-isomorphic respectively to $\mathfrak{A}_1/\overset{1}{K_0}$ and $\mathfrak{A}_2/\overset{2}{K_0}$. This allows us to specify the classical limit of φ as the *-homomorphism $\varphi_0 \colon \mathfrak{A}_1/\overset{1}{K_0} \to \mathfrak{A}_2/\overset{2}{K_0}$ defined by

$$\varphi_0([a]) = [\varphi(a)] \tag{5.7}$$

for any $a \in \mathfrak{A}_1$. Thus, we finally have an association $\varphi_1 \mapsto \varphi_0$ of structure-preserving and scaling maps between quantum models with structure-preserving maps between classical models.

Theorem 9 *The correspondence $\varphi_1 \mapsto \varphi_0$ is a bijection between the structure-preserving and scaling maps $\varphi_1 \colon \mathcal{W}(V_1, \sigma_1) \to \mathcal{W}(V_2, \sigma_2)$ and the structure-preserving maps $\varphi_0 \colon AP(V_1') \to AP(V_2')$.*[97]

Proof sketch of Theorem 9: The strategy of the proof is to show that there is an explicit inverse map $\varphi_0 \mapsto \varphi_1$ from structure-preserving maps $\varphi_0 \colon AP(V_1') \to AP(V_2')$ to structure-preserving scaling maps $\varphi_1 \colon \mathcal{W}(V_1, \sigma_1) \to \mathcal{W}(V_2, \sigma_2)$.

---

[97] Recent work of Belov-Kanel et al. (2021) has established a similar result for the corresponding polynomial algebras of generators of the Weyl unitaries.

To that end, suppose we are given a *-homomorphism $\varphi_0 \colon AP(V_1') \to AP(V_2')$ such that $\varphi_0[\Delta(V_1)] \subseteq \Delta(V_2)$ and φ_0 preserves the Poisson bracket.

One can show that there exists a character[98] $\chi \colon V_1 \to \mathbb{C}$ on the additive group V_1 and an additive symplectic transformation $T \colon V_1 \to V_2$ such that

$$\varphi_0(W_0(f_1)) = \chi(f_1)W_0(Tf_1) \tag{5.8}$$

for any $f_1 \in V_1$. We omit the proof, which follows from a combination of Gelfand duality (Theorem 3) – which allows one to identify elements of the classical Weyl algebra with continuous functions on the dual group of a vector space[99] – and Pontryagin duality[100] – which allows one to identify the second dual group of a vector space with the vector space itself.

We define the map $\varphi_\hbar \colon \mathcal{W}(V_1, \hbar\sigma_1) \to \mathcal{W}(V_2, \hbar\sigma_2)$ as the continuous linear extension of the assignment[101]

$$\varphi_\hbar(W_\hbar(f_1)) = \chi(f_1)W_\hbar(Tf_1) \tag{5.9}$$

for any $f_1 \in V_1$. One can show through direct calculation that $\varphi_\hbar$ is a *-homomorphism, that $\varphi_\hbar[Q_\hbar^W(\Delta(V_1))] \subseteq Q_\hbar^W(\Delta(V_2))$, and that when we set $\hbar = 1$, φ_1 is scaling and hence a structure-preserving map between the quantum models. Finally, one can check that the classical limit of φ_1 as defined through the preceding discussion is the map φ_0 we began with. □

The foregoing makes precise some claims that the algebraic and Poisson structure of classical physics is continuous with the structure of quantum physics. By showing that there is a correspondence between the structure-preserving maps in classical and quantum theories, we have shown a sense in which the structures of those theories correspond to each other. This structural continuity holds inherent and deep interest for philosophical issues such as how theory change bears on the interpretation of scientific theories.

Of course, the preceding argument applies only for a circumscribed collection of bosonic quantum theories with linear phase spaces whose physical quantities are represented by the Weyl algebra. But other results on the *functoriality* of more general quantization procedures (e.g., Bieliavsky and Gayral, 2015, p. 74 and p. 141) provide possible routes to generalize this result to a broader collection of quantum theories.

---

[98] Recall that a *character* on a group G is a map $\chi \colon G \to \mathbb{C}$ such that $\chi(g_1 + g_2) = \chi(g_1) \cdot \chi(g_2)$ and $\chi(e) = 1$ for $g_1, g_2 \in G$ and $e \in G$ the identity element.

[99] See Binz et al. (2004a, §IV.C) for this application of Gelfand duality to the classical Weyl algebra.

[100] See Rudin (1962, §1.7).

[101] See also Binz et al. (2004a, §III.D) and Binz et al. (2004b, §5.3).

Note also that structural realists have not necessarily singled out algebraic and Poisson structure in quantum physics as their target. For example, French (2012) identifies charge structure in quantum theories – as encoded in the representation structure of certain symmetry groups – as a focal point for structural realists in quantum field theories. This suggests the question of whether maps preserving charge structure in classical theories correspond with maps preserving charge structure in quantum theories. The work of Landsman (2003) contributes toward this question by showing how maps that preserve symmetry group representation structure in classical theories correspond to maps that preserve symmetry group representation structure in quantum theories. But there is much further work to be done here to assess the structural continuities between classical and quantum physics.

5.2 Analogies

Beyond issues in general philosophy of science like the realism debate, there are (of course) many interpretive questions that are specific to quantum theories. Some of these interpretive issues arise because the way in which the mathematical formalism of quantum theories is used to represent features of the physical world is not only difficult to understand, but also controversial; the way in which highly theoretical or mathematical aspects of the theory have physical meaning can often seem opaque. While philosophers of science have long been interested in how theoretical terms in highly mathematized scientific theories gain meaning, the purpose of this section is to take some inspiration from those general philosophical discussions to develop local strategies for interpreting quantum theories.

Our inspiration comes primarily from Hesse (1953, 1961, 1970), who emphasized the role of analogies between newly constructed and existing or already understood theories for interpreting and giving meaning to a mathematical formalism that is used to represent physical systems. Indeed, Hesse (1952) even discusses the analogies between classical and quantum physics as one of her examples of this pattern of analogical reasoning in scientific practice.[102] In this vein, we also mention Landsman (2017, see especially the Introduction), who focuses on the relation between classical and quantum physics as of central foundational importance in an attempt to make sense of aspects of interpreting quantum theory that he traces back to Bohr and Heisenberg. We hope to suggest a use of analogies between classical and quantum physics that

---

[102] For other contemporary work on analogies in the development of and interpretation of quantum theories, see Fraser and Koberinski (2016) and Fraser (2020a,b).

follows in the footsteps of these authors, but with somewhat more focused purview. We leave the interpretation of quantum theory in general (and, e.g., the measurement problem) outside the scope of our discussion, and instead, we hope to demonstrate that analogies can aid in the interpretation of specific physical quantities and mathematical concepts that appear in quantum theories.

To that end, we take up two case studies. First, we consider a continuity condition for states on the Weyl algebra called *regularity*. Second, we take up the issue of *inequivalent particle concepts* in quantum field theories.

5.2.1 Regularity

In this section, we argue that analogies between classical and quantum theories can aid in the interpretation of a typical continuity constraint on states on the Weyl algebra called the regularity condition. A state ω on the Weyl algebra $\mathcal{W}(V, \hbar\sigma)$ over a symplectic vector space (V, σ) is called *regular* if for each $f \in V$, the map

$$t \in \mathbb{R} \mapsto \omega(W_\hbar(tf)) \tag{5.10}$$

is a continuous function. In this section, we consider primarily the quantum theory of finitely many particles with phase space $V = \mathbb{R}^{2n}$. This is the setting in which Halvorson (2004) argues that the regularity condition is not justified. Considering nonregular states gives rise to the possibility of states with determinate position values or determinate momentum values – possibilities, which are usually thought to be precluded in quantum mechanics. We will argue that analogies to the classical case provide reasons to restrict attention to only regular states and thus to rule out determinate position and determinate momentum states as in standard quantum mechanics.[103] As we will see later, restricting attention to a collection of states is intimately related to choosing an appropriate algebra for quantization. In Section 6 we will see how restricting to regular states can also be thought of as a tool to compensate for unfortunate features of the Weyl algebra itself, and we will argue that one has reason to start from a different algebra. In this section, we will work only in the setting of the Weyl algebra to focus on interpreting the states themselves before we return to this issue later.

One reason in the quantum theory to restrict attention to regular states is that they allow one to define position and momentum operators, as follows. Given

---

[103] See also Teller (1979) and Halvorson (2001a).

a regular state ω on $\mathcal{W}(V, \hbar\sigma)$, the GNS representation π_ω for ω on the Hilbert space $\mathcal{H}_\omega$ has the feature that the one-parameter families of unitaries

$$t \in \mathbb{R} \mapsto \pi_\omega(W_\hbar(tf)) \tag{5.11}$$

for any fixed $f \in V$ are continuous in the weak operator topology on $\mathcal{B}(\mathcal{H}_\omega)$. It follows then from Stone's theorem (Reed and Simon, 1975, §VIII.4) that there is a (generally unbounded) self-adjoint operator $\Phi(f)$ on $\mathcal{H}_\omega$ generating the unitary family in the sense that

$$\pi_\omega(W_\hbar(tf)) = e^{it\Phi(f)} \tag{5.12}$$

for any $t \in \mathbb{R}$. In a field theory, the operators $\Phi(f)$ correspond to smeared field and field momentum operators. But in the theory of a single particle with phase space $V = \mathbb{R}^2$, the generators of the one-parameter unitary families

$$t \in \mathbb{R} \mapsto \pi_\omega(W_\hbar(0, t)) \qquad\qquad t \in \mathbb{R} \mapsto \pi_\omega(W_\hbar(t, 0)) \tag{5.13}$$

are the position operator Q and the momentum operator P, respectively, and this can similarly be generalized for the position and momentum operators of n particles. Conversely, failure of regularity implies that the unitary families in Eq. (5.11) are not all weak operator continuous and so either the position or the momentum operator will fail to be well-defined in the GNS representation of the nonregular state.

Let us now focus on the theory of finitely many particles with phase space $V = \mathbb{R}^{2n}$. In the remainder of this section, we will draw two analogies between the classical and quantum case. The first analogy concerns the continuity of the families of unitaries

$$t \in \mathbb{R} \mapsto W_\hbar(t(a, b)) \tag{5.14}$$

for fixed $(a, b) \in \mathbb{R}^{2n}$ at $\hbar > 0$ and $\hbar = 0$. The second analogy directly concerns the status of nonregular states with respect to the orthodox representation of states as density operators in the quantum theory and probability measures in the classical theory.

Let us first consider the families of unitaries in Eq. (5.14). One can draw analogies between the various ways in which these families of unitaries succeed and fail in being continuous. In both the classical algebra $AP(\mathbb{R}^{2n})$ and the quantum algebra $\mathcal{W}_\hbar(\mathbb{R}^{2n}, \hbar\sigma)$, these families of unitaries fail to be continuous in the abstract weak topology on the algebra. However, in the quantum theory, there is a different sense in which the unitary families are continuous. Namely, since the standard Schrödinger representation π_S is quasi-equivalent to the GNS representation of any regular state,[104] it follows that the unitary families

$$t \in \mathbb{R} \mapsto \pi_S(W_\hbar(tf)) \tag{5.15}$$

for $f \in \mathbb{R}^{2n}$ are continuous in the weak operator topology on $L^2(\mathbb{R}^n)$. Halvorson (2004) claims that assuming the regularity condition begs the question, which implies he thinks we have no prior reason to understand the unitary families in Eq. (5.14) as suitably continuous, and therefore no prior reason to use the Schrödinger representation. However, if we look at the classical case, there is an analogous sense in which the unitary families in Eq. (5.14) are continuous (in a topology different from the abstract weak topology). That is, the families of periodic functions

$$t \in \mathbb{R} \mapsto W_0(t(a,b)) \qquad W_0(t(a,b))(p,q) = e^{it(a \cdot p + b \cdot q)}$$

for $(p,q) \in \mathbb{R}^{2n}$ are indeed continuous with respect to the topology of pointwise convergence. Further, these families of functions are even continuous with respect to the topology generated by the semi-norms on $AP(\mathbb{R}^{2n})$ defined by

$$A \in AP(\mathbb{R}^{2n}) \mapsto \left| \int_{\mathbb{R}^{2n}} A \, d\mu \right| \tag{5.16}$$

for all countably additive Borel probability measures μ on $\mathbb{R}^{2n}$. In other words, for every countably additive Borel probability measure μ on $\mathbb{R}^{2n}$, the function

$$t \in \mathbb{R} \mapsto \left| \int_{\mathbb{R}^{2n}} W_0(t(a,b)) \, d\mu \right| \tag{5.17}$$

is continuous. So, looking at the classical case shows us that there is a natural sense in which the unitary families in Eq. (5.14) are continuous. In other words, in both the classical and quantum theories, there are analogous senses in which the unitary families in Eq. (5.14) are continuous (pointwise and in the weak operator topology) and fail to be continuous (in the abstract weak topology).[105] Insofar as the pointwise continuity of the classical unitary families is enough to justify considering their generators q and p, continuity in the weak operator topology of the Schrödinger representation should be enough as well.

But let us now consider directly the status of nonregular states in the classical and quantum theory. In the quantum theory, nonregular states cannot be represented by density operators in the Schrödinger representation of $\mathcal{W}_\hbar(\mathbb{R}^{2n}, \hbar\sigma)$. Again, the claim of Halvorson (2004) that assuming the regularity condition

---

[104] This is implied by the Stone–von Neumann theorem. See Mackey (1949), Summers (1999), and Petz (1990). See also Clifton and Halvorson (2001) and Ruetsche (2011) for philosophical discussion.

[105] See Feintzeig (2017) for more discussion.

begs the question implies that we have no prior reason to restrict attention to states that can be represented by density operators in the Schrödinger representation. The mathematical situation is analogous in the classical case, but in the more familiar classical setting it will become clear that we do have prior reason to restrict attention to regular states. On the classical algebra $AP(\mathbb{R}^{2n})$, any state represented by a countably additive Borel probability measure is a regular state. In particular, consider the state $\omega_{(p,q)}$ represented by (the delta function measure at) the point $(p,q) \in \mathbb{R}^{2n}$ and defined by the assignment

$$\omega_{(p,q)}(A) = A(p,q) \tag{5.18}$$

for all $A \in AP(\mathbb{R}^{2n})$. Each state $\omega_{(p,q)}$ is a pure state on $AP(\mathbb{R}^{2n})$. Recall that the Gelfand theorem (Theorem 3) implies that $AP(\mathbb{R}^{2n})$ is *-isomorphic to the algebra $C(\mathcal{P}(AP(\mathbb{R}^{2n})))$ of all continuous functions on its pure state space. But we have just established that there is a canonical injection

$$(p,q) \in \mathbb{R}^{2n} \mapsto \omega_{(p,q)} \in \mathcal{P}(AP(\mathbb{R}^{2n})). \tag{5.19}$$

Clearly, for each state of the form $\omega_{(p,q)}$, the function

$$t \in \mathbb{R} \mapsto \omega_{(p,q)}(W_0(t(a,b))) = e^{it(a \cdot p + b \cdot q)} \tag{5.20}$$

is continuous, and hence $\omega_{(p,q)}$ is regular. Moreover, it follows that any probability measure on $\mathcal{P}(AP(\mathbb{R}^{2n}))$ that is a convex combination of states of the form $\omega_{(p,q)}$ must regular, and hence every probability measure on $\mathcal{P}(AP(\mathbb{R}^{2n}))$ with support on the states of the form $\omega_{(p,q)}$ is regular. Since the Riesz–Markov theorem (Theorem 1) implies that every state on $AP(\mathbb{R}^{2n})$ is represented by a probability measure on $\mathcal{P}(AP(\mathbb{R}^{2n}))$, it follows that nonregular states must assign nonzero measure to the remainder $\mathcal{P}(AP(\mathbb{R}^{2n})) \setminus \mathbb{R}^{2n}$. Thus, nonregular states on the classical algebra $AP(\mathbb{R}^{2n})$ are sometimes called "states at infinity" because they have support outside the phase space $\mathbb{R}^{2n}$ assumed in the first place.[106] This is analogous to the way that nonregular states on $\mathcal{W}(\mathbb{R}^{2n}, \hbar\sigma)$ in the quantum theory lie outside the Hilbert space of the Schrödinger representation, which provides the orthodox state space of the quantum theory. Since in the classical theory there is apparently prior reason to restrict to states with support on $\mathbb{R}^{2n}$ – because this was the phase space assumed in the construction of the theory – we claim the regularity condition is justified. Insofar as the quantum theory is analogous to the classical one, the regularity condition will be justified there as well.[107]

---

[106] See also Feintzeig (2018b).
[107] See also Feintzeig et al. (2019) and Feintzeig and Weatherall (2019) for further discussion of regular states.

5.2.2 Inequivalent Particle Concepts

In this section, we demonstrate how analogies between classical and quantum theories can elucidate the philosophical discussion around inequivalent particle concepts in quantum field theories. We use the Klein–Gordon theory as our primary example. The space $V = C_c^\infty(\mathbb{R}^3) \oplus C_c^\infty(\mathbb{R}^3)$ consists of test functions for initial data for a real-valued scalar field with the symplectic form

$$\sigma((f,g),(\tilde{f},\tilde{g})) = \int_{\mathbb{R}^3} f(x)\tilde{g}(x) - \tilde{f}(x)g(x)d^3x \qquad (5.21)$$

for $(f,g),(\tilde{f},\tilde{g}) \in V$, as in Ex. 25. In this case, it is well known that the Weyl algebra has inequivalent Hilbert space representations. These Hilbert space representations are on Fock spaces composed of particle states corresponding to excitations from distinct vacuum states. We will consider the (standard) Minkowski vacuum state and the Rindler vacuum state, which each generate inequivalent Fock spaces and so inequivalent particle notions.[108] Our task in this section is to understand the significance of these inequivalent particle notions.

The Minkowski vacuum state is the lowest energy state for the Klein–Gordon field quantized in inertial coordinates. The Minkowski Fock space is constructed from the self-adjoint operator μ_M on $C_c^\infty(\mathbb{R}^3)$ and complex structure J_M on V defined by[109]

$$J_M(f,g) := (-\mu_M^{-1}g, \mu_M f) \qquad \mu_M := (m^2 - \nabla^2)^{1/2} \qquad (5.22)$$

for all $(f,g) \in V$. This then defines a complex inner product α_M on V by

$$\alpha_M(F,G) := \sigma(F, J_M G) + i\sigma(F,G) \qquad (5.23)$$

for all $F, G \in V$. These structures J_M and α_M on V are the unique ones that commute with time evolution by the Klein–Gordon equation in inertial coordinates. Letting $\mathcal{H}_M := \overline{V}^{\alpha_M}$ denote the Hilbert space completion of V with respect to α_M, this implies that the Klein–Gordon dynamics in inertial coordinates lifts to a unitary dynamics on the Minkowski Fock space $\mathcal{F}(\mathcal{H}_M)$. In fact, we know that $\mathcal{F}(\mathcal{H}_M)$ is the representing Hilbert space for the GNS representation of the state ω_M on $\mathcal{W}(V, \hbar\sigma)$ defined by

$$\omega_M(W_\hbar(F)) := e^{-\frac{\hbar}{4}\alpha_M(F,F)}, \qquad (5.24)$$

---

[108] See Kay (1985) for details on the construction of these Fock space representations. See also Kay and Wald (1991) for generalizations to curved spacetimes.

[109] Given a positive self-adjoint operator A, $A^{1/2}$ denotes the positive operator such that $A^{1/2}A^{1/2} = A$, whose existence and uniqueness is guaranteed by Reed and Simon (Theorem VIII.15 1975, p. 278).

for all $F \in V$, which serves as the vacuum state for the Minkowski Fock space and is invariant under time evolution in inertial coordinates.

On the other hand, the Rindler vacuum state is the lowest energy state for the Klein–Gordon field in accelerated Rindler coordinates. With respect to inertial Minkowski coordinates (t, x, y, z), the Rindler coordinates (T, X, y, z) are defined on the right Rindler wedge region

$$R := \{(t, x, y, z) \mid x \geq t\} \tag{5.25}$$

by the relations

$$t = e^X \sinh(T) \qquad\qquad x = e^X \cosh(T). \tag{5.26}$$

In these coordinates, Lorentz boosts generate the Rindler time translations $(T, X, y, z) \mapsto (T + T', X, y, z)$. As above, we consider an initial value formulation for the right Rindler wedge with initial data on the spacelike surface

$$\tilde{R} := \{(x, y, z) \mid x \geq 0\}. \tag{5.27}$$

The Rindler Fock space is constructed from the self-adjoint operator μ_R and complex structure J_R acting on $V(\tilde{R}) := C_c^\infty(\tilde{R}) \oplus C_c^\infty(\tilde{R})$ defined by

$$J_R(f, g) := (-\mu_R^{-1} g, \mu_R f) \qquad \mu_R := \left(e^X \left(m^2 - \frac{\partial^2}{\partial y^2} - \frac{\partial^2}{\partial z^2} \right) - \frac{\partial^2}{\partial X^2} \right)^{1/2} \tag{5.28}$$

for all $(f, g) \in V(R)$. This then defines a complex inner product α_R on $V(R)$ by

$$\alpha_R(F, G) := \sigma(F, J_R G) + i\sigma(F, G) \tag{5.29}$$

for all $F, G \in V(R)$. These structures J_R and α_R on $V(\tilde{R})$ are the unique ones that commute with time evolution by the Klein–Gordon equation in Rindler coordinates. Letting $\mathcal{H}_R := \overline{V(\tilde{R})}^{\alpha_R}$ denote the Hilbert space completion of $V(\tilde{R})$ with respect to α_R, this implies that the Klein–Gordon dynamics in Rindler coordinates lifts to a unitary dynamics on the Rindler Fock space $\mathcal{F}(\mathcal{H}_R)$. In fact, we know that $\mathcal{F}(\mathcal{H}_R)$ is the representing Hilbert space for the GNS representation of the state ω_R on $\mathcal{W}(V(\tilde{R}), \hbar\sigma)$ defined by

$$\omega_R(W_\hbar(F)) := e^{-\frac{\hbar}{4}\alpha_R(F, F)} \tag{5.30}$$

for all $F \in V(\tilde{R})$, which serves as the vacuum state for the Rindler Fock space and is invariant under time evolution in Rindler coordinates.

The first interpretive problem raised by these inequivalent particle notions is that the Minkowski and Rindler Fock spaces appear to provide competing and incommensurable frameworks for the theory of the free Klein–Gordon field.[110]

---

[110] See also Ruetsche (2002, 2003, 2011) and Clifton and Halvorson (2001).

The Minkowski Fock space $\mathcal{F}(\mathcal{H}_M)$ fails to contain any density operator state reproducing the expectation values of the Rindler vacuum ω_R, and the Rindler Fock space $\mathcal{F}(\mathcal{H}_R)$ fails to contain any density operator state reproducing the expectation values of the Minkowski vacuum ω_M. Further, the collections of states represented by density operators in the Minkowski Fock space $\mathcal{F}(\mathcal{H}_M)$ is disjoint from the collection of states represented by density operators in the Rindler Fock space $\mathcal{F}(\mathcal{H}_R)$. The Minkowski and Rindler representations, taken as entire theories of the free Klein–Gordon field, then disagree on what are the physically possible states. As Ruetsche (2011, chapters 9–10) puts it, this pushes us – if we want a unified mathematical framework for the theory – to either choose between the Minkowski and Rindler Fock space representations, or else reject the conservative dogma that one requires an irreducible Hilbert space representation of the physical quantities in a quantum theory at all. But Ruetsche argues that neither option is viable, claiming we should reject the desire for a unified mathematical framework for quantum field theories to begin with.

Such a surprising conclusion deserves further thought. Quantum theories certainly have forced us to accept many counterintuitive ideas, but we should do so with care. Since the appearance of inequivalent Hilbert space representations is unfamiliar in even standard nonrelativistic quantum mechanics for particle systems, we suggest trying to better understand these mathematical aspects of the theory through more familiar routes. Indeed, we will argue that drawing analogies to the classical field theory we quantized gives us a better handle on both the role of inequivalent Hilbert space representations and the possible interpretations of the inequivalent particle notions in the free Klein–Gordon theory. We will argue that analogies between the classical and quantum field theories lead to the conclusions that (i) Ruetsche is correct that we should not choose one or the other Fock space representation as the formalism for the entire quantum field theory, but (ii) we are not forced to follow Ruetsche in the rejection of a unified mathematical framework for the quantum field theory. We will conclude by investigating how the classical limit provides tools for a positive understanding of the inequivalent particle concepts in the Minkowski and Rindler Fock space representations.

Recall Ruetsche's reason that we should not focus on one or the other irreducible Fock space representation is that neither can represent all of the physically possible states of the theory. If this is correct, then we should understand each representation to correspond to some subset of the space of physically possible states. Indeed, this understanding of Hilbert space representations as subsets of physically possible states is supported by analogies with other physical theories. First, in the classical theory, we can represent the physical quantities of the Klein–Gordon field by the algebra $AP(V')$, which is the classical limit of the Weyl algebra $\mathcal{W}(V, \hbar\sigma)$. Each point in the phase space V' is a (possibly distributional) field configuration (π, φ), which determines a pure state $\omega_{(\pi,\varphi)}$ on $AP(V')$ by the continuous linear extension of the assignment

$$\omega_{(\pi,\varphi)}(W_0(f,g)) := W_0(f,g)(\pi, \varphi) = e^{i \int \pi(x)f(x) + \varphi(x)g(x)} \tag{5.31}$$

for all $(f,g) \in V$. But the GNS representation of any pure state on a commutative algebra corresponds to a one-dimensional Hilbert space (Kadison and Ringrose, 1997, pp. 744–747). So the GNS representation of the classical algebra $AP(V')$ for the state $\omega_{(\pi,\varphi)}$ corresponds to the representing Hilbert space $\mathcal{H} = \mathbb{C}$ with $\mathcal{B}(\mathcal{H}) = \mathbb{C}$. The representation is then given by the continuous linear extension of the assignment

$$W_0(f,g) \in AP(V') \mapsto \omega_{(\pi,\varphi)}(W_0(f,g)) \in \mathbb{C} = \mathcal{B}(\mathcal{H}) \tag{5.32}$$

for all $(f,g) \in V$. It follows immediately that the GNS representations for any two distinct field configurations are inequivalent. Thus, since in the classical case we take it as uncontroversial that there is more than one physically possible state in the theory, it follows that the irreducible representations corresponding to the pure states determined by field configurations correspond to only a small part of the theory. In the classical theory it is clear that one should not understand inequivalent Hilbert space representations as competing theories, but rather as parts of a larger theory.[111]

We also mention as an aside that one can use another quantum theory as an analogy to reach the same conclusion. Suppose one considers the quantum theory of a charged particle moving in a configuration space Q with an external Yang–Mills force field with symmetry group G. As in Ex. 23 the physical quantities of this theory are represented by the C*-algebra $\mathcal{K}(L^2(Q)) \otimes C^*(G)$. The algebra $\mathcal{K}(L^2(Q))$ of compact operators has a unique Hilbert space representation, but $C^*(G)$ has inequivalent representations corresponding to the unitary representations of G. Thus, the algebra of physical quantities has inequivalent representations corresponding to inequivalent representations of G. These are superselection sectors corresponding to different total charge values for the particle. Each irreducible Hilbert space representation contains only states with the same value for the total charge. But since multiple charge values are physically possible, none of these Hilbert spaces contains all of the physically possible states of the theory. In this case, it is again clear that one should not understand the inequivalent Hilbert

---

[111] See also Feintzeig (2015, 2016).

space representations as competing theories, but rather as parts of a larger theory.[112]

Now let us suppose we accept that we should not choose between inequivalent Hilbert space representations like the Minkowski and Rindler Fock space representations as if they were competing theories of the free Klein–Gordon field. The apparent alternative is to take the algebraic structure of the physical quantities that all representations share in common as the mathematical structure of the physical theory. But Ruetsche argues against the viability of this approach. She claims one cannot take an abstract C*-algebra to completely specify the quantum theory because there are physically significant quantities that do not belong to the abstract C*-algebra – in this case the Weyl algebra.[113] For example, no bounded function of the number operators belongs to the abstract Weyl algebra. Ruetsche claims that we require the extra structure of a Hilbert space to recover quantities like (bounded functions of) the number operator – which she dubs *parochial observables* – as limits in the weak operator topology. For this reason, she argues we cannot jettison Hilbert space representations altogether.

We will argue, however, that looking at the analogous limiting quantities in the classical theory shows that there is a way to recover parochial observables from the algebraic structure without a Hilbert space representation. For example, let us consider the quantities we obtain as weak operator limits in Hilbert space representations of the classical algebra $AP(V')$. Recall that $AP(V')$ is *-isomorphic to $C(\hat{V})$, the algebra of continuous functions on the topological space $\hat{V} \cong \mathcal{P}(AP(V'))$ of all characters on the additive group V, which corresponds to the pure state space of $AP(V')$ (Binz et al., 2004a). Any probability measure μ on $\hat{V}$ defines a Hilbert space representation of $C(\hat{V})$ on $L^2(\hat{V}, \mu)$ acting by left multiplication, defined explicitly as

$$ f \in C(\hat{V}) \mapsto M_f \in \mathcal{B}(L^2(\hat{V}, \mu)) \qquad (M_f \psi)(x) := f(x)\psi(x) \qquad (5.33) $$

for all $f \in C(\hat{V})$, $\psi \in L^2(\hat{V}, \mu)$, and $x \in \hat{V}$. Distinct probability measures in general correspond to inequivalent Hilbert space representations. In this case, the parochial observables for the representation on $L^2(\hat{V}, \mu)$ are weak operator limits of the quantities in $C(\hat{V})$, which correspond to left multiplication by bounded and measurable functions on $\hat{V}$. Taking weak operator limits in a Hilbert space representation thus allows us to recover even discontinuous physical quantities;

---

just as in the quantum theory, weak operator limits produce more physical quantities than belong to the Weyl algebra.

However, such discontinuous physical quantities can be recovered in the classical theory without resort to the structure of any Hilbert space representation. Every bounded and measurable function on $\hat{V}$ is a limit of continuous functions with respect to the (abstract) weak topology on $C(\hat{V})$, or even the topology of pointwise convergence on $\hat{V}$, neither of which require a Hilbert space representation for their definition.[114] In fact, Feintzeig (2018c) argues that this stands in analogy to the quantum case where every parochial observable can be obtained as a limit of functions in the abstract algebra with respect to the (abstract) weak topology. Thus, the analogy to the classical case suggests that the appearance of inequivalent Hilbert space representations does motivate a move to the abstract algebraic setting.

But neither analogy to the classical case just discussed (inequivalent Hilbert space representations or parochial observables) sheds much positive light on the distinct particle notions appearing in the Minkowski Fock space and the Rindler Fock space. Each of these Fock spaces comes with its own system of number operators. We claim that drawing analogies to the classical theory can also shed some light on the significance of these particle notions. Browning et al. (2020) show, using the strict quantization of the Weyl algebera, that one can take the classical $\hbar \to 0$ limit of both the Minkowski and Rindler total number operators $N_\hbar^M$ and $N_\hbar^R$ to find the classical quantities N_0^M and N_0^R, respectively, represented by real-valued functions on the phase space V'. The classical limit of the Minkowski number operator takes the form

$$N_0^M(\pi,\varphi) = \frac{1}{2} \int_{\mathbb{R}^3} (\mu_M^{1/2}\varphi)^2 + (\mu_M^{-1/2}\pi)^2 \qquad (5.34)$$

for all $\pi,\varphi \in C_c^\infty(\mathbb{R}^3)$, while the classical limit of the Rindler number operator takes the form

$$N_0^R(\pi,\varphi) = \frac{1}{2} \int_{\tilde{R}} (\mu_R^{1/2}\varphi)^2 + (\mu_R^{-1/2}e^x\pi)^2 \qquad (5.35)$$

for all $\pi,\varphi \in C_c^\infty(\tilde{R})$. Since these quantities approximate the theoretical structure of the number operators in the quantum theory, we can understand each classical number operator to represent the same physical quantity as the corresponding quantum number operator. With this understanding, we see that in the classical theory the Minkowski and Rindler total number quantities are just two different quantities associated with classical field configurations. This might

---

[114] See Appendix B of Feintzeig (2018c) for further discussion of the parochial observables in classical theories.

also assuage some of the worries described in Ruetsche (2011) if we do not have reason to take either particle concept as fundamental, even in the classical theory.

Finally, we mention that the particle concepts that emerge in the classical $\hbar \to 0$ limit may help with further foundational issues. To take just one illustration, some philosophers (Malament, 1996; Halvorson and Clifton, 2002) have argued that quantum field theories do not admit a concept of *localizable* particles. But Feintzeig et al. (2021) argue that localizable particles might be seen as emergent on the classical level by demonstrating senses in which the classical Minkowski number operator N_0^M can be understood locally. For example, on one simple interpretation N_0^M is localizable because it can be written as the integral of a density function, and integrating that density function over any local region gives rise to a quantity that can be understood as the number of particles in that local region.[115]

Thus, we conclude that analogies to the classical limit of quantum theories can aid in the interpretation of inequivalent particle concepts. Of course, this open-ended discussion leaves room for much future work interpreting particles in quantum field theory through analogies with the classical number operators. For example, since the quantities N_0^R and N_0^M take continuous values, one needs to say more to understand how they can come to approximate particle-like structures within classical continuous field theories.

6 Theory Construction

Quantization is often conceived of as a tool for constructing a quantum theory from a corresponding classical theory. Since this capacity of quantization has played only a cursory role in the preceding sections, we now devote this section to the topic.

We begin by noting that, although the questions of how quantum theories are constructed and why they should be constructed via those methods have a clear philosophical importance, questions of this type have largely been ignored by philosophers of science. This is partly due to the fact that philosophers of science have tended to focus on issues in the *context of justification*, while the use of quantization in theory construction belongs to the *context of discovery*. The well-known distinction between the context of discovery and the context of justification traces back as far as, for example, Reichenbach (1938) and Popper (1959). The context of discovery refers to scientific activities during the creative process of arriving at new ideas, concepts, or theories; the context of

---

[115] See also Wallace (2001) for further discussion of emergent localizable particles with some (distinct) analogies to classical field theories.

justification refers to the ways in which those ideas, concepts, or theories are validated through experiments or other means. Many of the classic topics of investigation in philosophy of science – for example, confirmation, empiricism, explanation – are concerned primarily with the context of justification, seemingly implying that there is less of philosophical interest in the context of discovery. Indeed, one way to motivate this is with the intuitive thought that we can only give descriptive accounts of discovery, while we can give normative accounts of justification, thus making it an appropriate target for philosophical inquiry.

One can, of course, find some philosophical work on the topic of scientific discovery. For example, Laudan (1980), Zahar (1983), and Nickles (1985) discuss the possibility of a philosophical or normative account of the discovery process – often referred to as a "logic of discovery." One can similarly find many historical case studies that cover the inception of a scientific theory. We mention, in particular, the work of Hesse (1961) on the implications of the history of physics for theory construction. Finally, we mention the influential work of Post (1971), who argues for ways in which a correspondence principle might be used quite generally in the construction of scientific theories.

In this section we will focus on the justification for quantization in particular as a method of theory construction, thus investigating just one particular instance of the philosophical issues in the context of discovery. What concerns us here is when we should use quantization as a method for constructing quantum theories, and the reasons why. We will first explore possible ways in which quantization might be justified as a heuristic procedure through the reductive explanations it gives rise to. Then we will consider a recent proposal that we should construct quantum theories without quantizing classical theories at all.

6.1 Heuristics for the Choice of Algebra

We will focus here on just one example of a way in which quantization can inform the construction of a quantum theory. As we have seen previously in §2.4, there are many different C*-algebras one might use to implement the canonical commutation relations for the same system.[116] For example, the physical quantities of a system of finitely many particles with the phase space $\mathbb{R}^{2n}$ might be represented by the Weyl algebra, the resolvent algebra, the compact operators, or even all the bounded operators on $L^2(\mathbb{R}^n)$. Each of these algebras has a different state space and hence corresponds to a different starting

---

[116] See also Feintzeig (2018a).

point concerning what states are physically possible in the kinematics of the theory. We will attempt to illustrate how quantization might aid in the choice of algebra for a particular physical system based on our conception of what states are physically possible in the corresponding classical theory. This shows that we can provide reasons to justify, or at least make plausible, some of the choices made in the construction of new theories.

The reasoning we pursue for the choice of algebra is that quantum theories should be able to explain the success of their classical predecessors, and so the two should be related by an intertheoretic reduction. Moreover, we understand this to be the stronger requirement for theoretical reduction discussed in §4.1 – that the quantum theory should explain the theoretical structure of its corresponding classical theory. The reason for this strong requirement is that it is an *empirical phenomenon* that scientists using the old classical theory formulated with a certain mathematical structure were successful at a variety of scientific tasks. We take it this phenomenon deserves explanation, and such an explanation can often be found based on how the old theory approximates the new one.

One aspect of the theoretical structure of the old theory that should be explained is the space of classical physical states. Since the old classical theory was successful with its state space, our desideratum is to show that the collection of physically possible classical states can be explained by the new quantum theory. One way to achieve such an explanation is to show that all and only the states deemed physically possible by the classical theory can be recovered as approximations to states deemed physically possible by the newly constructed quantum theory.[117] Indeed, this requirement is nontrivial and we will show that in two examples of types of physical systems it gives rise to guidance on the algebra to use in constructing a quantum theory – guidance that agrees with the traditional understanding of the quantum theory.

As a preliminary, we recall the methods outlined in §3 for analyzing classical limits of states. Suppose we are given a strict quantization of a Poisson algebra $\mathcal{P}$ with fiber algebras $(\mathfrak{A}_\hbar)_{\hbar \in [0,1]}$ and quantization maps $(Q_\hbar)_{\hbar \in [0,1]}$. We say that a family of states $\{\omega_\hbar\}_{\hbar \in [0,1]}$, where each $\omega_\hbar$ is a state on $\mathfrak{A}_\hbar$, is a *continuous field of states* if, for each $A \in \mathcal{P}$, the function $\hbar \mapsto \omega_\hbar(Q_\hbar(A))$ is continuous.[118] In a continuous field of states, ω_0 is understood as the classical limit of the states $\omega_\hbar$ for $\hbar > 0$, and the defining condition requires that the expectation

---

[117] See Feintzeig (2022).

[118] Actually, the notion of a continuous field of states can be defined directly in terms of the underlying continuous bundle of C*-algebras, and so does not rely on the particular quantization map.

values determined by ω_0 approximate those determined by the states $\omega_\hbar$ arbitrarily well in the limit $\hbar \to 0$. With this notion of the classical limit for states in mind, we turn to two example systems to illustrate our heuristic for theory construction in action.

6.1.1 Regularity Revisited

The first system we consider is our standard case of a system of finitely many particles with phase space $\mathbb{R}^{2n}$. In this case, physically possible classical states are represented by countably additive Borel probability measures on $\mathbb{R}^{2n}$, the pure states of which correspond to points in $\mathbb{R}^{2n}$. So one aim of a reductive explanation of the classical theory is to show that all and only these classical states can be recovered as classical limits of quantum states. But suppose one constructs the quantum theory for such a system with the Weyl algebra $\mathcal{W}(\mathbb{R}^{2n}, \hbar\sigma)$. The classical limit of the Weyl algebra is the algebra of almost periodic functions $AP(\mathbb{R}^{2n})$, which allows many "states at infinity" that cannot be represented as probability measures on $\mathbb{R}^{2n}$. Indeed, we identified in §5.2.1 analogies between such classical "states at infinity" and nonregular states on the Weyl algebra. Moreover, we will now show that the classical limit of a state on the Weyl algebra can be represented as a countably additive Borel probability measure on $\mathbb{R}^{2n}$ if and only if the original state on the Weyl algebra was regular.

We need one preliminary to state the result precisely. A state ω_0 on $AP(\mathbb{R}^{2n})$ determines a countably additive Borel probability measure μ_{ω_0} on the space $\mathcal{P}(AP(\mathbb{R}^{2n}))$ of all pure states on $AP(\mathbb{R}^{2n})$ such that

$$\omega_0(A) = \int_{\mathcal{P}(AP(\mathbb{R}^{2n}))} A \, d\mu_{\omega_0} \tag{6.1}$$

for all $A \in AP(\mathbb{R}^{2n})$. There is a canonical injection $k \colon \mathbb{R}^{2n} \to \mathcal{P}(AP(\mathbb{R}^{2n}))$ defined by

$$k(x)(A) := A(x) \tag{6.2}$$

for all $A \in AP(\mathbb{R}^{2n})$ and $x \in \mathbb{R}^{2n}$. But the remainder $\mathcal{P}(AP(\mathbb{R}^{2n})) \setminus k[\mathbb{R}^{2n}]$ is nonempty. When we say ω_0 can be represented by a (countably additive Borel) probability measure on $\mathbb{R}^{2n}$, this corresponds to the condition $\mu_{\omega_0}(\mathcal{P}(AP(\mathbb{R}^{2n})) \setminus k[\mathbb{R}^{2n}]) = 0$. Now we have the following result from Feintzeig (2018b).

Theorem 10 *A state ω_0 on $AP(\mathbb{R}^{2n})$ satisfies $\mu_{\omega_0}(\mathcal{P}(AP(\mathbb{R}^{2n})) \setminus k[\mathbb{R}^{2n}]) = 0$ if and only if there is a continuous field of states $\{\omega_\hbar\}_{\hbar \in [0,1]}$ relative to the Berezin quantization map (Eq. (3.24)) with each $\omega_\hbar$ a regular state on the Weyl algebra for $\hbar > 0$.*

Proof Sketch of Theorem 10: Suppose first that $\{\omega_\hbar\}_{\hbar \in [0,1]}$ is a continuous field of states and each $\omega_\hbar$ for $\hbar > 0$ is regular. It follows from Theorem 2 of Feintzeig (2018b) that the state $\omega_\hbar$ on $\mathcal{W}(\mathbb{R}^{2n}, \hbar\sigma)$ is regular if and only if $\omega_\hbar \circ Q_\hbar^B$ extends to a state on $AP(\mathbb{R}^{2n})$ such that $\mu_{\omega_\hbar \circ Q_\hbar^B}(\mathcal{P}(AP(\mathbb{R}^{2n})) \setminus k[\mathbb{R}^{2n}]) = 0$. Then the definition of a continuous field of states implies that for all $A \in AP(\mathbb{R}^{2n})$,

$$
\begin{aligned}
\int_{\mathcal{P}(AP(\mathbb{R}^{2n}))} A \, d\mu_{\omega_0} &= \omega_0(A) \\
&= \lim_{\hbar \to 0} \omega_\hbar \circ Q_\hbar^B(A) \\
&= \lim_{\hbar \to 0} \int_{\mathcal{P}(AP(\mathbb{R}^{2n}))} A \, d\mu_{\omega_\hbar \circ Q_\hbar^B} \\
&= \lim_{\hbar \to 0} \int_{k[\mathbb{R}^{2n}]} A \, d\mu_{\omega_\hbar \circ Q_\hbar^B},
\end{aligned}
\tag{6.3}
$$

which implies $\mu_{\omega_0}(\mathcal{P}(AP(\mathbb{R}^{2n})) \setminus k[\mathbb{R}^{2n}]) = 0$. The converse follows directly from the result of Waldmann (2010) (see also Kaschek et al. (2009)). $\qquad \square$

Thus, in order to explain the success of the classical state space, we need to use a different algebra in the quantum theory whose states reproduce all and only the countably additive Borel probability measures in the classical limit.[119] This can be accomplished by employing the algebra $\mathcal{K}(L^2(\mathbb{R}^n))$ of compact operators, whose classical limit is the algebra $C_0(\mathbb{R}^{2n})$ of continuous functions vanishing at infinity. The latter algebra $C_0(\mathbb{R}^{2n})$ allows for only classical states that can be represented as countably additive Borel probability measures on $\mathbb{R}^{2n}$. Thus, the heuristic requirement that our newly constructed quantum theory explain why the states of the old classical theory were good approximations leads to the result that we should employ $\mathcal{K}(L^2(\mathbb{R}^n))$ rather than $\mathcal{W}(\mathbb{R}^{2n}, \hbar\sigma)$ in the quantum theory. Using the algebra $\mathcal{K}(L^2(\mathbb{R}^n))$ also corresponds to ruling out nonregular states in the quantum theory since all states on $\mathcal{K}(L^2(\mathbb{R}^n))$ can be represented by density operators on $L^2(\mathbb{R}^n)$. In this case, the restriction to quantum states whose classical limits are physical states gives rise to the standard formulation of the kinematics of quantum theory with regular states on the Hilbert space $L^2(\mathbb{R}^n)$.

Given that we have just argued against the use of the Weyl algebra in quantum theories, one might wonder whether any result analogous to Theorem 10 holds for other algebras. For example, suppose one employs the entire

---

[119] Indeed, others have also identified the choice of algebra as the central issue for ruling out nonregular states. See Grundling (1997) and Grundling and Neeb (2009).

algebra $\mathcal{B}(L^2(\mathbb{R}^n))$ of bounded operators, which allows for many *singular states* (Kadison and Ringrose, 1997, p. 723) that – just as nonregular states on the Weyl algebra – cannot be understood as density operators in the standard Schrödinger representation on $L^2(\mathbb{R}^n)$. Since the Berezin quantization map $Q_\hbar^B$ (Eq. (3.12)) is a continuous map from $C_c^\infty(\mathbb{R}^{2n})$ to the compact operators $\mathcal{K}(L^2(\mathbb{R}^n))$, and since the weak operator closure of the compact operators is the algebra of all bounded operators, it follows that there is a unique continuous extension of $Q_\hbar^B$ to a map from the algebra $C_b(\mathbb{R}^{2n})$ of continuous, bounded functions to $\mathcal{B}(L^2(\mathbb{R}^n))$, and moreover this map agrees with the Schrödinger representation of the Berezin quantization (Eq. (3.24)) of $AP(\mathbb{R}^{2n})$ (Browning et al., 2020, Prop. 1). Hence, we can think of the algebra $\mathcal{B}(L^2(\mathbb{R}^n))$ of bounded operators as obtained by quantizing the algebra $C_b(\mathbb{R}^{2n})$ of continuous, bounded functions, and we can consider continuous fields of states on the continuous bundle of C*-algebras defined by this quantization. But since a state ω on $\mathcal{B}(L^2(\mathbb{R}^n))$ is nonsingular just in case $\omega \circ \pi_S$ is regular, we have the following immediate corollary. In the statement of the result, we now let $k\colon \mathbb{R}^{2n} \to \mathcal{P}(C_b(\mathbb{R}^{2n}))$ denote the canonical injection defined in exact analogy with Eq. (6.2).

Corollary 1 *A state ω_0 on $C_b(\mathbb{R}^{2n})$ satisfies $\mu_{\omega_0}(\mathcal{P}(C_b(\mathbb{R}^{2n})) \setminus k[\mathbb{R}^{2n}]) = 0$ if and only if there is a continuous field of states $\{\omega_\hbar\}_{\hbar \in [0,1]}$ relative to the extension of the Berezin quantization map (Eq. (3.12)) to $C_b(\mathbb{R}^{2n})$ with each $\omega_\hbar$ a nonsingular state on $\mathcal{B}(L^2(\mathbb{R}^n))$ for $\hbar > 0$.*

Proof Sketch of Corollary 1: Suppose $\{\omega_\hbar\}_{\hbar \in [0,1]}$ is a continuous field of states, where each state $\omega_\hbar$ for $\hbar > 0$ is nonsingular on $\mathcal{B}(L^2(\mathbb{R}^n))$. This holds if and only if for each $\hbar > 0$, $\omega_\hbar \circ \pi_S$ is a regular state on $\mathcal{W}(\mathbb{R}^{2n}, \hbar\sigma)$, and hence, by Theorem 10, the restriction of the state ω_0 to $AP(\mathbb{R}^{2n})$ is representable by a countably additive probability measure on $\mathbb{R}^{2n}$, which holds if and only if $\mu_{\omega_0}(\mathcal{P}(C_b(\mathbb{R}^{2n})) \setminus k[\mathbb{R}^{2n}]) = 0$ by Prop. 2 of Feintzeig (2018b). $\qquad\square$

This establishes that the main result of this section does not depend on the use of the Weyl algebra, which one might take as an objectionable starting point. Indeed, we have shown that, to explain the classical state space, one should employ in the quantum theory only states that can be represented as density operators in the Schrödinger representation. This robust result guides our choice of algebra, providing reason to restrict attention to either regular states on $\mathcal{W}(\mathbb{R}^{2n}, \hbar\sigma)$ or nonsingular states on $\mathcal{B}(L^2(\mathbb{R}^n))$. Ultimately, both can be accomplished by using the algebra $\mathcal{K}(L^2(\mathbb{R}^n))$ of compact operators.

6.1.2 Symmetry Invariance

The second system we consider is that of a charged particle moving in an external Yang–Mills gauge field. This includes, for example, a charged particle moving in an external electromagnetic field. The classical theory of such a system is represented by a principal bundle $P \to Q$ over a base space given by a Riemannian manifold Q representing the configuration space, with typical fiber given by a compact unimodular Lie group G representing the symmetry group of the gauge field.[120] In the classical theory, one can take the phase space to be the cotangent bundle T^*P directly. However, the theory is typically reformulated using the dynamical symmetries encoded in G by instead employing the Marsden–Weinsten reduction procedure to arrive at the so-called universal phase space $(T^*P)/G$.[121] Here, we denote the action of G on P by R_g for each $g \in G$, and the action of G on T^*P is given by the pull-back R_g^* for each $g \in G$. The universal phase space $(T^*P)/G$ is a Poisson manifold that decomposes into symplectic leaves, each corresponding to a collection of states with a fixed charge, or what might be thought of as a charge sector of the classical theory.[122] So we take physical states in the classical theory to be countably additive Borel probability measures on the universal phase space $(T^*P)/G$.

In this case, one can quantize T^*P directly using the Weyl quantization maps to obtain the algebra $\mathcal{K}(L^2(P))$ using the Weyl quantization map in Eq. (3.34). However, the classical limit of this algebra is $C_0(T^*P)$, whose states are countably additive Borel probability measures on T^*P. The states ω on $C_0(T^*P)$ that are considered physically possible in the classical theory are ones that are invariant under the action of G in the sense that $\omega(f) = \omega(R_g^*f)$ for each $g \in G$. Such states are in one-to-one correspondence with the states on the universal phase space on the algebra $C_0((T^*P)/G)$. One can see this by defining for each $f \in C_0(T^*P)$, a corresponding function $\pi_0(f) \in C_0((T^*P)/G)$ by

$$\pi_0(f)(xG) := \int_G f(R_g x) \, dg, \tag{6.4}$$

where dg is the Haar measure on G. The map π_0 is a surjection, and for any state ω^G on $C_0((T^*P)/G)$, the state ω on $C_0(T^*P)$ defined by

$$\omega(f) := \omega^G(\pi_0(f)) \tag{6.5}$$

will be G-invariant because $\pi_0(R_g^*f) = \pi_0(f)$ for any $g \in G$. One can check that the assignment $\omega^G \mapsto \omega$ is a one-to-one correspondence between the states on $C_0((T^*P)/G)$ and the G-invariant states on $C_0(T^*P)$.

---

[120] See, for example, Sternberg (1977) for the classical Hamiltonian formulation of such a system.
[121] See Marsden and Weinstein (1974) and Weinstein (1978).
[122] See chapter III.2 of Landsman (1998a).

Indeed, these G-invariant classical states are analogous to quantum states represented by G-invariant density operators on $L^2(P)$. More precisely, there is a unitary representation of G on $L^2(P)$ by

$$(U(g)\psi)(x) := \psi(R_g x) \tag{6.6}$$

for $g \in G$, $\psi \in L^2(P)$, and $x \in P$. A density operator ρ on $L^2(P)$ defines a state ω_ρ on $\mathcal{K}(L^2(P))$ by

$$\omega_\rho(A) := Tr(\rho A) \tag{6.7}$$

for $A \in \mathcal{K}(L^2(P))$, and every state on $\mathcal{K}(L^2(P))$ corresponds to a unique density operator. We say that a density operator ρ is G-invariant, and hence the state ω_ρ is G-invariant, if

$$\rho = U(g)\rho U(g)^* \tag{6.8}$$

for every $g \in G$. Analogous to the classical case, G-invariant states on $\mathcal{K}(L^2(P))$ are in one-to-one correspondence with states on the algebra $\mathcal{K}(L^2(P))^G$ of G-invariant compact operators. Here, a compact operator A is called G-invariant just in case

$$A = U(g)AU(g)^* \tag{6.9}$$

for every $g \in G$. One can define for each $A \in \mathcal{K}(L^2(P))$ a corresponding operator $\pi_1(A) \in \mathcal{K}(L^2(P))$ by

$$(\pi_1(A)\psi)(x) := \int_G (U(g)AU(g)^*\psi)(x)\, dg$$

for $\psi \in L^2(P)$ and $x \in P$. The map π_1 is a surjection, and for any state ω^G on $\mathcal{K}(L^2(P))^G$, the state ω on $\mathcal{K}(L^2(P))$ defined by

$$\omega(A) := \omega^G(\pi_1(A)) \tag{6.10}$$

will be G-invariant because $\pi_1(U(g)AU(g)^*) = \pi_1(A)$ for any $g \in G$.

Moreover, the classical limit of a state on $\mathcal{K}(L^2(P))$ is a G-invariant classical state on $C_0(T^*P)$ if and only if the original state on $\mathcal{K}(L^2(P))$ was G-invariant. More specifically, Browning and Feintzeig (2020) prove the following result for the Weyl quantization given in Eqs. (3.34)–(3.35).

Theorem 11 *A state ω_0 on $C_0(T^*P)$ is G-invariant if and only if there is a continuous field of states $\{\omega_\hbar\}_{\hbar \in [0,1]}$ relative to the Weyl quantization map (Eqs. (3.34)–(3.35)) with each $\omega_\hbar$ for $\hbar > 0$ a G-invariant state on $\mathcal{K}(L^2(P))$.*

Proof Sketch of Theorem 11: One can check from the definition of the Weyl quantization map $Q_\hbar^W$ on T^*P and by repeated application of the Fubini–Tonelli theorem that for any $f \in \mathcal{P}$,

$$Q_\hbar^W(\pi_0(f)) = \pi_1 \circ Q_\hbar^W(f). \tag{6.11}$$

Further, it follows from the preceding discussion that (i) the state ω_0 is G-invariant if and only if $\omega_0 \circ \pi_0 = \omega_0$, and (ii) each state $\omega_\hbar$ is G-invariant if and only if $\omega_\hbar \circ \pi_1 = \omega_\hbar$.

So if each $\omega_\hbar$ for $\hbar > 0$ is G-invariant, then we have that, for each $f \in \mathcal{P}$,

$$
\begin{aligned}
\omega_0(f) &= \lim_{\hbar \to 0} \omega_\hbar(Q_\hbar^W(f)) \\
&= \lim_{\hbar \to 0} \omega_\hbar \circ \pi_1(Q_\hbar^W(f)) \\
&= \lim_{\hbar \to 0} \omega_\hbar(Q_\hbar^W(\pi_0(f))) \\
&= \omega_0(\pi_0(f)).
\end{aligned}
\tag{6.12}
$$

This implies $\omega_0 = \omega_0 \circ \pi_0$, and hence ω_0 is G-invariant.

Conversely, suppose ω_0 is a G-invariant state on $C_0(T^*P)$. It follows from Landsman (1993, p. 105) that there is a continuous field of states $\{\omega_\hbar\}_{\hbar \in [0,1]}$ converging to ω_0. Hence, the continuous field of states $\{\omega_\hbar \circ \pi_1\}$ satisfies

$$
\begin{aligned}
\lim_{\hbar \to 0} \omega_\hbar \circ \pi_1(Q_\hbar(f)) &= \lim_{\hbar \to 0} \omega_\hbar(Q_\hbar(\pi_0(f)) \\
&= \omega_0(\pi_0(f)) \\
&= \omega_0(f)
\end{aligned}
\tag{6.13}
$$

for each $f \in C_0(T^*P)$. Since $\omega_\hbar \circ \pi_1$ is a G-invariant state on $\mathcal{K}(L^2(P))$, we have constructed a continuous field of states converging to ω_0 that is G-invariant for each $\hbar > 0$. $\qquad\square$

Thus, in order to explain the success of the classical state space we need to use a different algebra in the quantum theory whose states reproduce all and only the G-invariant states – or countably additive Borel probability measures on $(T^*P)/G$ – in the classical limit. This can be accomplished by employing the algebra $\mathcal{K}(L^2(P))^G$ of G-invariant compact operators on $L^2(P)$, whose classical limit is the algebra $C_0(T^*P)^G$ of G-invariant continuous functions vanishing at infinity on T^*P. The classical algebra $C_0(T^*P)^G$ is *-isomorphic to the algebra $C_0((T^*P)/G)$ of continuous functions vanishing at infinity on the universal phase space $(T^*P)/G$, and thus its states are in one-to-one correspondence with the countably additive Borel probability measures on $(T^*P)/G$, which we have taken to be the physically possible classical states.

On the other hand, the quantum algebra $\mathcal{K}(L^2(P))^G$ is *-isomorphic to $\mathcal{K}(L^2(Q)) \otimes C^*(G)$ (Landsman, 1993, p. 108). States on the quantum algebra can be understood as density operators in representations, but now each representation of the algebra consists in the (unique) representation of $\mathcal{K}(L^2(Q))$ along with a unitary representation of G. The chosen unitary representation of G then corresponds to a fixed value for the (total) charge that acts as its generator. In this case, the restriction to quantum states whose classical limits are physical states in the universal phase space thus gives rise to the standard formulation of the kinematics of the quantum theory with wave functions and density operators in $L^2(Q)$ indexed by charge values associated with a unitary representation of G.

Thus the requirement that a quantum theory produces a reductive explanation of the structure of the state space in its corresponding classical theory through the $\hbar \to 0$ limit is successful in our two examples. In both cases, this motivates the choice of a C*-algebra of physical quantities for the quantum theory with precisely those states whose classical limits are physical states. We have seen that such a choice reproduces the typical setting for the kinematics of the quantum theory of finitely many particles with regular states and the quantum theory of a particle in an external gauge field with gauge-invariant states.

6.2 Do We Need Quantization at All?

The previous section establishes that quantization can provide heuristic procedures that lend plausibility, if not justification, to the quantum theories we construct. However, some have taken the view that quantization is not necessary and perhaps should not be pursued for constructing quantum field theories. For example, such a view is suggested by some of the canonical works in the tradition of algebraic quantum field theory. In his seminal book, Haag (1992) writes,

> We have interpreted the elements of $\mathcal{A}(O)$ as representing physical operations performable within O and we have seen that this interpretation tells us how to compute collision cross sections once the correspondence $O \to \mathcal{A}(O)$ is known. This suggests that the *net of algebras* $\mathcal{A}$ [...] constitutes the intrinsic mathematical description of the theory. (Haag, 1992, p. 105)

This seems to echo ideas found in earlier papers, for example, Haag and Kastler (1964), and one also finds a similar sentiment in Haag (1993) as he discusses how aspects of particular models of quantum field theory are encoded in the structure of state spaces of local algebras. Haag and Ojima (1996) even explicitly contrast the algebraic approach with standard quantization procedures for field theories:

> In the process of quantization the point-like fields become singular objects and, as a consequence, the non linear field equations cannot be carried over as they stand. Remedies to overcome these difficulties have been devised (infinite renormalization, etc.). They are pragmatically successful but even if one is willing to believe that one ultimately arrives at a well-defined theory in this way, the path by which one arrives at this is not very transparent. The formal Lagrangean is a heuristic crutch, but an indispensable one as long as no other way of defining the theory is known. So it seems desirable to develop an alternative approach which aims at defining the theory directly in the setting of relativistic quantum theory. (Haag and Ojima, 1996, p. 386)

The idea seems to be that the net of algebras will encode the entire structure of the theory without requiring any quantization procedure. For example, the energy and momentum of the system can be recovered as the generators of Poincaré transformations in a suitable Hilbert space representation. And the charge structure along with the gauge symmetries that are essential for specifying a particular field theory can be associated with the structure of the local algebras as well.[123] So this suggests the following question: why should we follow a quantization procedure at all, if we could construct quantum theories without them?

One problem with this point of view is of course the difficulties constructing interacting quantum field theories in the C*-algebraic framework. But there is a recent proposal for constructing interacting quantum theories without quantization, which has not yet been analyzed by philosophers. We take up this proposal as a case study in the remainder of this section.

Buchholz and Fredenhagen (2020a,b) construct a C*-algebra for a quantum theory given only a classical Lagrangian and without employing a quantization procedure. They write of their construction:

> Let us emphasize that we are not introducing "quantization rules" for the underlying classical theory. The classical theory primarily serves to describe the localization properties of the S-operators and to indicate which particular observable we have in mind, without trying to specify its concrete quantum realization. Thus, in accord with the doctrine of Niels Bohr, we are using "common language" in order to describe observables and operations relating to the quantum world. (Buchholz and Fredenhagen, 2020a, p. 948)

The idea seems to be that while the specification of and interpretation of the quantum theory may involve reference to structures from a corresponding classical theory, that classical information does not play the

---

[123] For this, see chapter IV of Haag (1992) and the discussion of DHR theory contained in Halvorson (2007).

central role in theory construction that one often has in mind in speaking of quantization.[124]

There are actually two potentially novel features of the Buchholz and Fredenhagen view. The first feature is the one just mentioned that the net of C*-algebras is constructed from a classical Lagrangian without a quantization procedure, as we will review in some detail next. The second feature is that Buchholz and Fredenhagen interpret the elements of their algebra as "operations" rather than "observables" or "physical quantities." This actually harkens back to some of the language used early on in the algebraic tradition (e.g., Haag and Kastler, 1964, pp. 849–850). However, Buchholz and Fredenhagen suggest specifically that their operations correspond to "the impact of temporary perturbations of the dynamics on the underlying states" (2020b, p. 151), which makes them somewhat different from the kinds of measurement operations considered in other works. In fact, the contrast can be seen because Buchholz and Fredenhagen claim that[125]

> one can recover from the operations the standard interpretation of states in terms of "primitive observables," which have a consistent statistical interpretation in accordance with basic principles of quantum physics. (Buchholz and Fredenhagen, 2020a, p. 963)

This suggests the interpretation of algebra elements as operations is somehow more basic than the interpretation as observables, which is at least prima facie a second difference with the quantization methods we have discussed that typically treat C*-algebras of physical quantities. In this section, we will consider the proposal outlined by Buchholz and Fredenhagen for the purpose of analyzing the extent to which these two features distinguish their approach to constructing quantum theories from quantization, as discussed here.

To define Buchholz and Fredenhagen's *dynamical C*-algebra*, we require some preliminaries. We work in a theory with configuration space E. As our primary example, we will take E to be the space of all global configurations of a scalar field on Minkowski spacetime $\mathcal{M}$. In this case, we set $E := C^\infty(\mathcal{M})$. We must work with a collection $\mathcal{F}$ of real-valued functionals on E satisfying the following conditions:

---

[124] One might notice that both Buchholz and Fredenhagen (2020a) and Landsman (2017) relate their ideas to those of Bohr, while the latter claims they are connected to quantization and the former authors reject such a connection.

[125] This claim relies on the results of Buchholz and Størmer (2015). It is discussed further in §6 of Buchholz and Fredenhagen (2020b).

(i) Given any field configuration $\phi_0 \in E$ and any functional $F \in \mathcal{F}$, there is a functional $F^{\phi_0} \in \mathcal{F}$ defined by

$$F^{\phi_0}(\phi) = F(\phi + \phi_0). \tag{6.14}$$

(ii) For any field configurations $\phi_1, \phi_2, \phi_3 \in E$, if the supports of ϕ_1 and ϕ_2 are disjoint, then

$$F(\phi_1 + \phi_2 + \phi_3) = F(\phi_1 + \phi_3) - F(\phi_3) + F(\phi_2 + \phi_3). \tag{6.15}$$

Further, each functional F is associated with a spacetime support in $\mathcal{M}$ corresponding to the region on which F depends on the values of the fields.[126] If $F \in \mathcal{F}$ takes the form

$$F(\phi) = \sum_{n=0}^{N} \int_M g_n(x)\phi(x)^n \, dx \tag{6.16}$$

for $\phi \in E$, where the test functions g_n belong to $C_c^\infty(\mathcal{M})$, then the support of F is the union of the supports of the test functions g_n.

In addition to the functionals in $\mathcal{F}$, we also start with the structure of a classical Lagrangian. For any test function $f \in C_c^\infty(\mathcal{M})$, we assume we have a smeared Lagrangian $L(f): E \to \mathbb{R}$. For example, if the Lagrangian comes from a Lagrangian density $\mathcal{L}(\phi): \mathcal{M} \to \mathbb{R}$, then the smeared Lagrangian has the form

$$L(f)(\phi) = \int_M (\mathcal{L}(\phi))(x)f(x) \, dx \tag{6.17}$$

for $f \in C_c^\infty(\mathcal{M})$ and $\phi \in E$. The Lagrangian L defines a family of *relative Lagrangians* $\delta L(\phi_0): E \to \mathbb{R}$ for $\phi_0 \in C_c^\infty(\mathcal{M})$ by

$$\delta L(\phi_0)(\phi) := L(f_0)^{\phi_0}(\phi) - L(f_0)(\phi) \tag{6.18}$$

for $\phi \in E$, where the test function $f_0 \in C_c^\infty(\mathcal{M})$ is such that $f_0(x) = 1$ whenever x is in the support of ϕ_0 (in which case the definition of $\delta L(\phi_0)$ is independent of the choice of f_0).

Given the data encoded in the configuration space E, the space of functionals $\mathcal{F}$, and the Lagrangian L, Buchholz and Fredenhagen define a C*-algebra $\mathcal{A}_L$ as follows.

Definition 8 The *dynamical C*-algebra* $\mathcal{A}_L$ for E and $\mathcal{F}$ is generated from elements $S(F)$ for $F \in \mathcal{F}$ called *relative S-operators* satisfying the relations

---

[126] See Brunetti et al. (2019, p. 529) and Rejzner (2016) for these properties as they appear in the introduction of spacetime structure to algebraic classical field theory and perturbative quantum field theory. See also Rejzner (2019).

(i) For $\phi_0 \in C_c^\infty(\mathcal{M})$ and $F \in \mathcal{F}$,

$$S(F)S(\delta L(\phi_0)) = S(F^{\phi_0} + \delta L(\phi_0)) = S(\delta L(\phi_0))S(F). \qquad (6.19)$$

(ii) For $F_1, F_2, F_3 \in \mathcal{F}$ with the support of F_1 later than that of F_2,

$$S(F_1 + F_2 + F_3) = S(F_1 + F_3)S(F_3)^{-1}S(F_2 + F_3). \qquad (6.20)$$

The set of all elements $S(F)$ for $F \in \mathcal{F}$ form a group. $\mathcal{A}_L$ is the completion of the collection of complex linear combinations of the group elements $S(F)$ in a suitable C*-norm.

The defining conditions of the dynamical C*-algebra $\mathcal{A}_L$ are relations discovered for relative S-operators in perturbative quantum field theory.[127] Eq. (6.19) encodes the dynamical structure while Eq. (6.20) encodes a causality condition. We mention also that $\mathcal{A}_L$ contains a proper subalgebra satisfying the Weyl commutation relations, and so it is strictly larger than the Weyl algebra.

As for the interpretation of the elements of the dynamical algebra $\mathcal{A}_L$, we will quote Buchholz and Fredenhagen at length. For the first difference with quantization procedures, they write:

> [I]n contrast to the ideas of Haag, we have used a bottom-up approach, where the operators, generating the algebra, are labelled from the outset by physical quantities, such as "field," "interaction potential," "relative action," *etc.* In the spirit of Bohr, we regard these labels as notions in the framework of classical field theory, which are merely used to describe which kind of objects in the quantum world we have in mind. There is no *a priori* quantization rule for them. The realization of the corresponding operators in the mathematical setting of quantum theory is fixed by the notion of causality, involving time ordering; the actual results depend on the chosen Lagrangean. (Buchholz and Fredenhagen, 2020a, p. 963)

As we see, here they explicitly assert their algebra is not obtained by quantization. And for the second difference with quantization procedures, they write:

> The present results seem to suggest, however, a change of paradigm in the interpretation of the Haag-Kastler framework. Originally, it was proposed to interpret the selfadjoint elements of the local algebras as observables. Yet such an interpretation does not fit well with our construction.[...]

---

[127] See also Brunetti et al. (2009), Fredenhagen and Rejzner (2015), and Fredenhagen and Rejzner (2016) for more on the origin of these defining relations for the relative S-operators.

> [W]e propose to interpret the unitary operators [$S(F)$] as operations, describing the impact of measurements of the conceived observables on quantum states in the given spacetime regions. (Buchholz and Fredenhagen, 2020a, p. 963)

Here again they assert that algebra elements do not have the (typical) interpretation as physical quantities or observables, on their view.

We suggest, however, that it is an important open question whether the dynamical algebra $\mathcal{A}_L$ can be obtained by a quantization procedure, even if it is agreed that the original authors did not construct the algebra in that way. To see this possibility, notice that we have other examples for which this is true. The resolvent algebra (Ex. 13) was originally constructed with only quantum theories in mind (Buchholz and Grundling, 2008) as a way to encode the canonical commutation relations while avoiding technical features that prevent the representation of a large class of dynamics on the Weyl algebra (Fannes and Verbeure, 1974). So one might have thought, likewise, that constructing a quantum theory using the resolvent algebra avoids quantization. However, van Nuland (2019) showed later that there is a corresponding classical resolvent algebra (Ex. 10) from which the quantum resolvent algebra can be obtained by strict quantization, and which serves as the classical $\hbar \to 0$ limit of the quantum resolvent algebra. So it does not follow in general that when a new algebraic structure is proposed for a quantum theory directly that such a structure cannot be obtained by quantization from the classical theory. We suggest that it would be of interest to discover whether it is possible to formulate a classical algebra that serves as the classical limit of the dynamical algebra $\mathcal{A}_L$ and is related by strict quantization.

Settling this question may even aid in the interpretation of the dynamical algebra itself in the quantum realm. For example, it is not clear from the discussion of Buchholz and Fredenhagen whether the interpretation of the S-operators $S(F)$ in the dynamical algebra as operations is forced upon us somehow or rather added to the formalism by the authors. Of course, this interpretation seems to have been important for the authors' own thinking in developing their framework and results. But one might approach their interpretive claim with some skepticism because it has for a long time been recognized that algebra elements play a dual role as either observables or generators of transformations, including possible dynamics.[128] The Weyl unitaries, for example, may either be interpreted as symmetry transformations in phase space corresponding to "shifts" in position or momentum, or alternatively as physical quantities that are

---

[128] See Zalamea (2018) for a general discussion or Swanson (2019) for similar remarks in the context of CPT symmetry.

periodic functions in position and momentum. So it is not clear that Buchholz and Fredenhagen's interpretation of algebra elements as operations is genuinely distinct from an interpretation as observables. However, if one had a classical algebra corresponding to $\mathcal{A}_L$ through quantization, then one could use this as a tool to assess the novelty of Buchholz and Fredenhagen's interpretation, and whether the elements $S(F)$ might be given a dual interpretation as observables or quantities associated with the system.

Without a concrete idea of what algebra might play the role of the classical limit of $\mathcal{A}_L$, there is not much more we can say. We take these suggestions concerning the Buchholz–Fredenhagen dynamical algebra, however, to provide just one more illustration of how attention to the classical–quantum correspondence may inform ongoing work in the foundations of quantum theories. There are outstanding interesting questions concerning the classical limits of quantum theories, which promise to yield fruit for the interpretation of those theories, and of quantization in general.

We close with one final remark about this section's title question. We must admit that in some general sense, quantization is not *needed* for theory construction. After all, if by quantization one means strict deformation quantization, then the originators of quantum theory certainly did not employ quantization. Moreover, with a broad enough notion of possibility, it is always *possible* that one could guess at a quantum theory without any quantization procedure at all. But this does not give us any reason to do away with quantization; instead, it only redirects us to the question of whether we *should* employ quantization.

We claim that there are some reasons to employ quantization in the construction of quantum theories. First, insofar as one wants to pursue a conservative strategy to theory construction, quantization provides such a recipe. That is, quantization successfully describes the relationship between many examples of classical theories and the quantum theories constructed from them. If one has the goal in theory construction to only construct another model within the current framework for quantum physics, and not to go beyond the framework itself, then there is some inductive evidence that quantization can accomplish this goal. Second, the tools of quantization can aid the interpretation of a newly constructed quantum formalism by relating them back to a more familiar and better understood classical theory. This also creates the possibility of providing heuristics for altering the new theoretical structure *in the process of theory construction* to fit with a desired interpretation or physical application, as discussed in this section. The preceding considerations are, of course, defeasible, and one must always check whether they are applicable in a given theoretical context. However, these are real virtues of quantization, and so if one wants to reject quantization as a method of constructing quantum theories, then one ought to provide some reason in favor of alternative methods.

7 Conclusion

The purpose of this Element has been to serve as an invitation for aspiring researchers interested in the classical–quantum correspondence. We hope to have demonstrated that the tools of quantization theory open up a host of both new interpretive questions about quantum theory, as well as new approaches to longstanding philosophical issues. Of course, the idea that philosophers should interpret scientific theories by looking to emergent structures – found by analyzing relationships between physical theories at different scales – is not new. In fact, such an approach has been advocated for recently in the interpretation of quantum field theories by, for example, Wallace (2001, 2006, 2011) and Williams (2018).[129] The classical limit is just one way to analyze emergent structures on different scales, while many other contemporary authors approach such issues with renormalization group methods for treating scaling limits. However, we note here that even if one is interested in other scaling limits, there is still reason to think that the tools developed for the classical limit will be helpful. Mathematical methods for renormalization in algebraic quantum field theory developed by Buchholz and Verch (1995, 1998) bear a close resemblance to the methods of quantization theory,[130] and although the precise relationship between these different scaling limits is still open, it is possible that comparing these different kinds of limiting procedures may illuminate the interpretive issues arising in these other areas.

Thus, not only does quantization theory provide tools for (i) interpreting quantum theories and the intertheory relation between classical and quantum physics and (ii) treating the classical–quantum correspondence as a case study for general issues in philosophy of science, but also (iii) the foundational issues broached in our discussion may have implications for the understanding of other kinds of scaling limits that are of interest to philosophers of physics. There is good reason for philosophers of physics to take up these issues in future research.

---

[129] See also, for example, Butterfield (2011a,b) and Batterman (2018) for this kind of thinking in statistical mechanics, or even Landsman and Reuvers (2013) for an application to the measurement problem.

[130] See also Buchholz (1996a,b).

Symbols

$\mathfrak{A}$	C*-algebra
$S(\mathfrak{A})$	state space
$\mathcal{P}(\mathfrak{A})$	pure state space
$C(X)$	continuous functions
$C_0(X)$	continuous functions vanishing at infinity
$C^\infty(X)$	smooth functions
$C_c^\infty(X)$	smooth, compactly supported functions
$L^2(X)$	square-integrable functions
$\mathcal{H}$	Hilbert space
$\mathcal{B}(\mathcal{H})$	bounded operators
$\mathcal{K}(\mathcal{H})$	compact operators
$\mathcal{F}(\mathcal{H})$	Fock space
$\langle \cdot, \cdot \rangle$	inner product
$[\cdot, \cdot]$	commutator
$\{\cdot, \cdot\}$	Poisson bracket

References

Alfsen, E. and Shultz, F. (2001). *State Spaces of Operator Algebras.* Birkhauser, Boston, MA.

Aliprantis, C. and Border, K. (1999). *Infinite Dimensional Analysis: A Hitchhiker's Guide.* Springer, Berlin.

Anzai, H. and Kakutani, S. (1943). Bohr compactifications of a locally compact abelian group I. *Proceedings of the Imperial Academy*, 19(9):476–480.

Ashtekar, A. and Isham, C. (1992). Inequivalent observable algebras: Another ambiguity in field quantisation. *Physics Letters B*, 274(3-4):393–398.

Baez, J., Segal, I., and Zhou, Z. (1992). *Introduction to Algebraic and Constructive Quantum Field Theory.* Princeton University Press, Princeton, NJ.

Barrett, T. W. (2020). Structure and equivalence. *Philosophy of Science*, 87(5):1184–1196.

Batterman, R. (1991). Chaos, quantization, and the correspondence principle. *Synthese*, 89(2):189–227.

Batterman, R. (1995). Theories between theories: Asymptotic limiting intertheoretic relations. *Synthese*, 103(2):171–201.

Batterman, R. (1997). "Into a mist": Asymptotic theories on a caustic. *Studies in the History and Philosophy of Modern Physics*, 28(3):395–413.

Batterman, R. (2002). *The Devil in the Details: Asymptotic Reasoning in Explanation, Reduction, and Emergence.* Oxford University Press, Oxford.

Batterman, R. (2005). Response to Belot's "Whose devil? Which details?" *Philosophy of Science*, 72(1):154–163.

Batterman, R. (2018). Autonomy of theories: An explanatory problem. *Nous*, 52(4):858–873.

Bayen, F., Flato, M., Fronsdal, C., Lichnerowicz, A., and Sternheimer, D. (1978a). Deformation theory and quantization. I. Deformations of symplectic structures. *Annals of Physics*, 111(1):61–110.

Bayen, F., Flato, M., Fronsdal, C., Lichnerowicz, A., and Sternheimer, D. (1978b). Deformation theory and quantization. II. Physical applications. *Annals of Physics*, 111(1):111–151.

Beller, M. (1999). *Quantum Dialogue.* The University of Chicago Press, Chicago.

Belot, G. (2005). Whose devil? Which details? *Philosophy of Science*, 72(1):128–153.

Belov-Kanel, A., Elishev, A., and Yu, J.-T. (2021). On automorphisms of the tame polynomial automorphism group in positive characteristic. https://arxiv.org/abs/2103.12784.

Berry, M. (1994). Asymptotics, singularities and the reduction of theories. In Skyrms, B., Prawitz, D., and Westerståhl, D. (eds.), *Logic, Methodology and Philosophy of Science, IX: Proceedings of the Ninth International Congress of Logic, Methodology, and Philosophy of Science, Uppsala, Sweden Aug 7–14, 1991*, pages 597–607.

Bieliavsky, P. and Gayral, V. (2015). *Deformation Quantization for Actions of Kählerian Lie Groups*, volume 236 of Memoirs of the American Mathematical Society. American Mathematical Society, Providence, RI.

Binz, E., Honegger, R., and Rieckers, A. (2004a). Construction and uniqueness of the C*-Weyl algebra over a general pre-symplectic space. *Journal of Mathematical Physics*, 45(7):2885–2907.

Binz, E., Honegger, R., and Rieckers, A. (2004b). Field-theoretic Weyl quantization as a strict and continuous deformation quantization. *Annales de l'Institut Henri Poincaré*, 5:327–346.

Bokulich, A. (2008). *Reexamining the Quantum-Classical Relation: Beyond Reductionism and Pluralism*. Cambridge University Press, Cambridge, UK.

Bordemann, M. and Waldmann, S. (1998). Formal GNS construction and states in deformation quantization. *Communications in Mathematical Physics*, 195:549–583.

Boyd, R. N. (1973). Realism, underdetermination, and a causal theory of evidence. *Noûs*, 7(1):1.

Boyd, R. (1989). What realism implies and what it does not. *Dialectica*, 43 (1–2):5–29.

Brading, K. and Landry, E. (2006). Scientific structuralism: Presentation and representation. *Philosophy of Science*, 73(5):571–581.

Bratteli, O. and Robinson, D. (1987). *Operator Algebras and Quantum Statistical Mechanics*, volume 1. Springer, New York.

Bratteli, O. and Robinson, D. (1996). *Operator Algebras and Quantum Statistical Mechanics*, volume 2. Springer, New York.

Browning, T. and Feintzeig, B. (2020). Classical limits of symmetry invariant states and the choice of algebra for quantization. *Letters in Mathematical Physics*, 110(7):1835–1860.

Browning, T., Feintzeig, B., Gates, R., Librande, J., and Soiffer, R. (2020). Classical limits of unbounded quantities by strict quantization. *Journal of Mathematical Physics*, 61(11):112305.

Brunetti, R., Dütsch, M., and Fredenhagen, K. (2009). Perturbative algebraic quantum field theory and the renormalization groups. *Advances in Theoretical and Mathematical Physics*, 13(5):1541–1599.

Brunetti, R., Fredenhagen, K., and Ribeiro, P. L. (2019). Algebraic structure of classical field theory: Kinematics and linearized dynamics for real scalar fields. *Communications in Mathematical Physics*, 368(2):519–584.

Buchholz, D. (1996a). Phase space properties of local observables and structure of scaling limits. *Annales de l'Institut Henri Poincaré*, 64(4):433–459.

Buchholz, D. (1996b). Quarks, gluons, colour: Facts or fiction. *Nuclear Physics B*, 469(1-2):333–353.

Buchholz, D. (2017). The resolvent algebra for oscillating lattice systems: Dynamics, ground and equilibrium states. *Communications in Mathematical Physics*, 353(2):691–716.

Buchholz, D. (2018). The resolvent algebra of non-relativistic Bose fields: Observables, dynamics and states. *Communications in Mathematical Physics*, 362(3):949–981.

Buchholz, D. and Fredenhagen, K. (2020a). A C*-algebraic approach to interacting quantum field theories. *Communications in Mathematical Physics*, 377(2):947–969.

Buchholz, D. and Fredenhagen, K. (2020b). Classical dynamics, arrow of time, and genesis of the Heisenberg commutation relations. *Expositiones Mathematicae*, 38(2):150–167.

Buchholz, D. and Grundling, H. (2008). The resolvent algebra: A new approach to canonical quantum systems. *Journal of Functional Analysis*, 254(11): 2725–2779.

Buchholz, D. and Grundling, H. (2015). Quantum systems and resolvent algebras. In Blanchard, B. and Fröhlich, J. (eds.), *The Message of Quantum Science: Attempts Towards a Synthesis*, pages 33–45. Springer, Berlin.

Buchholz, D. and Størmer, E. (2015). Superposition, transition probabilities and primitive observables in infinite quantum systems. *Communications in Mathematical Physics*, 339(1):309–325.

Buchholz, D. and Verch, R. (1995). Scaling algebras and renormalization group in algebraic quantum field theory. *Reviews in Mathematical Physics*, 7(8):1195.

Buchholz, D. and Verch, R. (1998). Scaling algebras and renormalization group in algebraic quantum field theory. II. Instructive examples. *Reviews in Mathematical Physics*, 10(6):775–800.

Bueno, O. (1999). What is structural empiricism? Scientific change in an empiricist setting. *Erkenntnis*, 50(1):59–85.

Butterfield, J. (2011a). Emergence, reduction, and supervenience: A varied landscape. *Foundations of Physics*, 41:920–959.

Butterfield, J. (2011b). Less is different: Emergence and reduction reconciled. *Foundations of Physics*, 41:1065–1135.

Callender, C. and Huggett, N. (eds.) (2001). *Physics Meets Philosophy at the Planck Scale*. Cambridge University Press, Cambridge, UK.

Cao, T. Y. (2003). Structural realism and the interpretation of quantum field theory. *Synthese*, 136(1):3–24.

Clifton, R. and Halvorson, H. (2001). Are Rindler quanta real? Inequivalent particle concepts in quantum field theory. *British Journal for the Philosophy of Science*, 52(3):417–470.

Crowther, K. (2021). Defining a crisis: The role of principles in the search for a theory of quantum gravity. *Synthese*, 198 (Suppl 14), 3489–3516.

Curd, M., Cover, J., and Pincock, C. (2013). *Philosophy of science: The central issues*. 2nd edition. W. W. Norton, New York.

Darrigol, O. (1992). *From c-Numbers to q-Numbers*. University of California Press, Berkeley.

Dewar, N. (2017). Interpretation and equivalence; or, equivalence and interpretation. Unpublished. http://philsci-archive.pitt.edu/13234/1/EI-2.pdf.

Dewar, N. (2019). Algebraic structuralism. *Philosophical Studies*, 176(7):1831–1854.

Dewar, N. (2022). *Structure and Equivalence*. Cambridge University Press, Cambridge, UK.

Dirac, P. (1930). *The Principles of Quantum Mechanics*. Oxford University Press, Oxford.

Dixmier, J. (1977). *C*-Algebras*. North Holland, New York.

Douglas, M. (2004). Report on the status of the Yang-Mills Millenium Prize Problem. www.claymath.org/sites/default/files/ym2.pdf.

Duncan, A. and Janssen, M. (2007a). On the verge of Umdeutung in Minnesota: Van Vleck and the correspondence principle. Part one. *Archive for History of Exact Sciences*, 61(6):553–624.

Duncan, A. and Janssen, M. (2007b). On the verge of Umdeutung in Minnesota: Van Vleck and the correspondence principle. Part two. *Archive for History of Exact Sciences*, 61(6):625–671.

Duncan, A. and Janssen, M. (2012). (Never) Mind your p's and q's: Von Neumann versus Jordan on the foundations of quantum theory. *The European Physical Journal H*, 38(2):175–259.

Duncan, A. and Janssen, M. (2019). *Constructing Quantum Mechanics: Volume 1: The Scaffold: 1900–1923*. Oxford University Press, Oxford.

Duncan, A. and Janssen, M. (2022). Quantization conditions, 1900–1927. In Freire, O. (ed.), *The Oxford Handbook of the History of Quantum Interpretations*, pages 76–94. Oxford University Press, Oxford.

Emch, G. (1972). *Algebraic Methods in Statistical Mechanics and Quantum Field Theory*. Wiley, New York.

Emch, G. (1983). Geometric dequantization and the correspondence problem. *International Journal of Theoretical Physics*, 22(5):397–420.

Fannes, M. and Verbeure, A. (1974). On the time evolution automorphisms of the CCR-algebra for quantum mechanics. *Communications in Mathematical Physics*, 35(3):257–264.

Fedosov, B. (1996). *Deformation Quantization and Index Theory*. Akademie Verlag, Berlin.

Feintzeig, B. (2015). On broken symmetries and classical systems. *Studies in the History and Philosophy of Modern Physics*, 52, Part B:267–273.

Feintzeig, B. (2016). Unitary inequivalence in classical systems. *Synthese*, 193(9):2685–2705.

Feintzeig, B. (2017). On theory construction in physics: Continuity from classical to quantum. *Erkenntnis*, 82(6):1195–1210.

Feintzeig, B. (2018a). On the choice of algebra for quantization. *Philosophy of Science*, 85(1):102–125.

Feintzeig, B. (2018b). The classical limit of a state on the Weyl algebra. *Journal of Mathematical Physics*, 59:112102.

Feintzeig, B. (2018c). Toward an understanding of parochial observables. *British Journal for the Philosophy of Science*, 69(1):161–191.

Feintzeig, B. (2022). Reductive explanation and the construction of quantum theories. *British Journal for Philosophy of Science*, 73(2):457–486.

Feintzeig, B., Manchak, J., Rosenstock, S., and Weatherall, J. (2019). Why be regular? Part I. *Studies in the History and Philosophy of Modern Physics*, 65:122–132.

Feintzeig, B. and Weatherall, J. (2019). Why be regular? Part II. *Studies in the History and Philosophy of Modern Physics*, 65:133–144.

Feintzeig, B. H. (2020). The classical limit as an approximation. *Philosophy of Science*, 87(4):612–539.

Feintzeig, B. H., Librande, J., and Soiffer, R. (2021). Localizable particles in the classical limit of quantum field theory. *Foundations of Physics*, 51(2):49.

Feyerabend, P. (1962). Explanation, reduction, and empiricism. In Feigl, H. and Maxwell, G. (eds.), *Minnesota Studies in the Philosophy of Science: Scientific Explanation, Space, and Time*, volume 3, pages 28–97. University of Minnesota Press, Minneapolis.

Fine, A. (1988). *The Shaky Game: Einstein, Realism, and the Quantum Theory*. University of Chicago Press, Chicago.

Forman, P. (1971). Weimar culture, causality, and quantum theory, 1918–1927: Adaptation by German physicists and mathematicians to a hostile intellectual environment. *Historical Studies in the Physical Sciences*, 3:1–115.

Fraser, D. (2020a). The development of renormalization group methods for particle physics: Formal analogies between classical statistical mechanics and quantum field theory. *Synthese*, 197:3027–3063.

Fraser, D. (2020b). The non-miraculous success of formal analogies in quantum theories. In French, S. and Saatsi, J., editors, *Scientific Realism and the Quantum*. Oxford University Press, Oxford.

Fraser, D. and Koberinski, A. (2016). The Higgs mechanism and superconductivity: A case study of formal analogies. *Studies in History and Philosophy of Science Part B: Studies in History and Philosophy of Modern Physics*, 55:72–91.

Fredenhagen, K. and Rejzner, K. (2015). Perturbative construction of models in algebraic quantum field theory. In Brunetti, R., Dappiaggi, C., Fredenhagen, K., and Yngvason, J., editors, *Advances in Algebraic Quantum Field Theory*, pages 31–74. Springer, Cham.

Fredenhagen, K. and Rejzner, K. (2016). Quantum field theory on curved spacetimes: Axiomatic framework and examples. *Journal of Mathematical Physics*, 57(3):031101.

French, S. (2012). Unitary inequivalence as a problem for structural realism. *Studies in History and Philosophy of Science Part B: Studies in History and Philosophy of Modern Physics*, 43(2):121–136.

French, S. and Ladyman, J. (2003). Remodelling structural realism: Quantum physics and the metaphysics of structure. *Synthese*, 136(1):31–56.

French, S. and Saatsi, J. (2006). Realism about structure: The semantic view and nonlinguistic representations. *Philosophy of Science*, 73(5): 548–559.

Frigg, R. and Votsis, I. (2011). Everything you always wanted to know about structural realism but were afraid to ask. *European Journal for Philosophy of Science*, 1(2):227–276.

Gamelin, T. (1969). *Uniform Algebras*. Prentice Hall, Englewood Cliffs, NJ.

Gotay, M. J. (1980). Functorial geometric quantization and Van Hove's theorem. *International Journal of Theoretical Physics*, 19(2):139–161.

Gotay, M. J. (1999). On the Groenewold–Van Hove problem for $\mathbb{R}^{2n}$. *Journal of Mathematical Physics*, 40(4):2107–2116.

Gracia-Bondía, J. M. (1992). Generalized Moyal quantization on homogeneous symplectic spaces. *Contemporary Mathematic*, 134:93–114.

Groenewold, H. (1946). On the principles of elementary quantum mechanics. *Physica*, 12(7):405–460.

Grundling, H. (1997). A group algebra for inductive limit groups: Continuity problems of the canonical commutation relations. *Acta Applicandae Mathematicae*, 46:107–145.

Grundling, H. and Neeb, K.-H. (2009). Full regularity for a C*-algebra of the Canonical Commutation Relations. *Reviews in Mathematical Physics*, 21(5):587–613.

Gutt, S. (1983). An explicit ⋆-product on the cotangent bundle of a Lie group. *Letters in Mathematical Physics*, 7(3):249–258.

Haag, R. (1992). *Local Quantum Physics*. Springer, Berlin.

Haag, R. (1993). Local quantum physics and models. *Communications in Mathematical Physics*, 155(1):199–204.

Haag, R. and Kastler, D. (1964). An algebraic approach to quantum field theory. *Journal of Mathematical Physics*, 5(7):848–861.

Haag, R. and Ojima, I. (1996). On the problem of defining a specific theory within the frame of local quantum physics. *Annales de L'Institut Henri Poincare-physique Theorique*, 64(4):385–393.

Halvorson, H. (2001a). On the nature of continuous physical quantities in classical and quantum mechanics. *Journal of Philosophical Logic*, 30:27–50.

Halvorson, H. (2001b). Reeh–Schlieder defeats Newton–Wigner: On alternative localization schemes in relativistic quantum field theory. *Philosophy of Science*, 68(1):111–133.

Halvorson, H. (2004). Complementarity of representations in quantum mechanics. *Studies in the History and Philosophy of Modern Physics*, 35(1): 45–56.

Halvorson, H. (2007). Algebraic quantum field theory. In Butterfield, J. and Earman, J., editors, *Handbook of the Philosophy of Physics*, volume 1, pages 731–864. Elsevier, New York.

Halvorson, H. (2016). Scientific theories. In Humphreys, P., editor, *The Oxford Handbook of Philosophy of Science*. Oxford University Press, Oxford.

Halvorson, H. and Clifton, R. (2002). No place for particles in relativistic quantum theories? *Philosophy of Science*, 69(1):1–28.

Hesse, M. (1952). Operational definition and analogy in physical theories. *British Journal for Philosophy of Science*, 2(8):281–294.

Hesse, M. (1953). Models in physics. *British Journal for Philosophy of Science*, 4(15):198–214.

Hesse, M. (1961). *Forces and Fields: The Concept of Action at a Distance in the History of Physics*. Dover, Mineola.

Hesse, M. (1970). *Models and Analogies in Science*. University of Notre Dame Press Notre Dame, IN.

Hewitt, E. (1953). Linear functionals on almost periodic functions. *American Mathematical Society*, 74(2):303–322.

Honegger, R. and Rieckers, A. (2005). Some continuous field quantizations, equivalent to the C*-Weyl quantization. *Publications of the Research Institute for Mathematical Sciences, Kyoto University*, 41(1):113–138.

Honegger, R., Rieckers, A., and Schlafer, L. (2008). Field-theoretic Weyl deformation quantization of enlarged Poisson algebras. *Symmetry, Integrability and Geometry: Methods and Applications*, 4:047–084.

Howard, D. (1986). What makes a classical concept classical? Toward a reconstruction of Niels Bohr's philosophy of physics. In *Niels Bohr and Contemporary Philosophy*, pages 201–229. Kluwer Academic Publishers, New York.

Hudetz, L. (2019). Definable categorical equivalence. *Philosophy of Science*, 86(1):47–75.

Jacobs, C. (2021). The coalescence approach to inequivalent representation: Pre-QM_∞ parallels. *The British Journal for the Philosophy of Science*. DOI: https://doi.org/10.1086/715108.

Jaffe, A. and Witten, E. (2000). Quantum Yang–Mills theory. www.claymath .org/sites/default/files/yangmills.pdf.

Jammer, M. (1966). *The Conceptual Development of Quantum Mechanics*. McGraw-Hill, New York.

Kadison, R. and Ringrose, J. (1997). *Fundamentals of the Theory of Operator Algebras*. American Mathematical Society, Providence, RI.

Kaschek, D., Neumaier, N., and Waldmann, S. (2009). Complete positivity of Rieffel's deformation quantization by actions of $\mathbb{R}^d$. *Journal of Noncommutative Geometry*, 3(3):361–375.

Kay, B. (1979). A uniqueness result in the Segal-Weinless approach to linear Bose fields. *Journal of Mathematical Physics*, 20(8):1712–3.

Kay, B. (1985). The double-wedge algebra for quantum fields on Schwarzschild and Minkowski spacetimes. *Communications in Mathematical Physics*, 100(1):57–81.

Kay, B. and Wald, R. (1991). Theorems on the uniqueness and thermal properties of stationary, nonsingular, quasifree states on spacetimes with a bifurcate killing horizon. *Physics Reports*, 207(2):49–136.

Kirchberg, E. and Wasserman, S. (1995). Operations on continuous bundles of C*-algebras. *Mathematische Annalen*, 303(4):677–697.

Kontsevich, M. (2003). Deformation quantization of Poisson manifolds. *Letters in Mathematical Physics*, 66(3):157–216.

Kuhn, T. (1970[1962]). *The Structure of Scientific Revolutions*. 2nd edition. University of Chicago Press, Chicago.

Kuhn, T. S. (1984). Revisiting Planck. *Historical Studies in the Physical Sciences*, 14(2):231–252.

Kuhn, T. S. (1987). *Black-Body Theory and the Quantum Discontinuity, 1894–1912*. The University of Chicago Press, Chicago.

Ladyman, J. (1998). What is structural realism? *Studies in History and Philosophy of Science Part A*, 29(3):409–424.

Landsman, N. P. (1990a). C*-algebraic quantization and the origin of topological quantum effects. *Letters in Mathematical Physics*, 20:11–18.

Landsman, N. P. (1990b). Quantization and superselection sectors I. Transformation group C*-algebras. *Reviews in Mathematical Physics*, 2(1): 45–72.

Landsman, N. P. (1993). Strict deformation quantization of a particle in external gravitational and Yang–Mills fields. *Journal of Geometry and Physics*, 12(2):93–132.

Landsman, N. P. (1998a). *Mathematical Topics between Classical and Quantum Mechanics*. Springer, New York.

Landsman, N. P. (1998b). Twisted Lie group C*-algebras as strict quantizations. *Letters in Mathematical Physics*, 46:181–188.

Landsman, N. P. (1999). Lie groupoid C*-algebras and Weyl quantization. *Communications in Mathematical Physics*, 206(2):367–381.

Landsman, N. P. (2003). Quantization as a functor. In Voronov, T., editor, *Quantization, Poisson Brackets and beyond*, pages 9–24. *Contemporary Mathematics*, 315, AMS. DOI: https://doi.org/10.1090/conm/315.

Landsman, N. P. (2007). Between classical and quantum. In Butterfield, J. and Earman, J., editors, *Handbook of the Philosophy of Physics*, volume 1, pages 417–553. Elsevier, New York.

Landsman, N. P. (2013). Spontaneous symmetry breaking in quantum systems: Emergence or reduction? *Studies in the History and Philosophy of Modern Physics*, 44(4):379–394.

Landsman, N. P. (2017). *Foundations of Quantum Theory: From Classical Concepts to Operator Algebras*. Springer, Cham.

Landsman, N. P. and Reuvers, R. (2013). A flea on Schrödinger's cat. *Foundations of Physics*, 43(3):373–407.

Laudan, L. (1980). Why was the logic of discovery abandoned? In Nickles, T., editor, *Scientific Discovery, Logic, and Rationality*, pages 173–183. Dordrecht Reidel, Dordrecht.

Laudan, L. (1981). A confutation of convergent realism. *Philosophy of Science*, 48(1):19–49.

Lee, R.-Y. (1976). On the C*-algebras of operator fields. *Indiana University Mathematics Journal*, 25(4):303.

Mackey, G. (1949). A theorem of Stone and von Neumann. *Duke Mathematical Journal*, 16(2):313–326.

Mackey, G. W. (1968). *Induced Representations of Groups and Quantum Mechanics*. W. A. Benjamin, Inc., New York.

Malament, D. (1996). In defense of dogma – why there cannot be a relativistic quantum mechanical theory of (localizable) particles. In Clifton, R., editor, *Perspectives on Quantum Reality*. Kluwer Academic Publishers, Amsterdam.

Manuceau, J., Sirugue, M., Testard, D., and Verbeure, A. (1974). The smallest C*-algebra for the canonical commutation relations. *Communications in Mathematical Physics*, 32:231–243.

Marsden, J. and Weinstein, M. (1974). Reduction of symplectic manifolds with symmetry. *Reports on Mathematical Physics*, 5(1):121–30.

Martinez, A. (2002). *An Introduction to Semiclassical and Microlocal Analysis*. Springer, New York.

Mazzucchi, S. (2009). *Mathematical Feynman Path Integrals and Their Applications*. World Scientific, Hackensack, NJ.

Moyal, J. E. (1949). Quantum mechanics as a statistical theory. *Mathematical Proceedings of the Cambridge Philosophical Society*, 45(1):99–124.

Murray, F. J. and von Neumann, J. (1936). On rings of operators. *The Annals of Mathematics*, 37(1):116.

Nagel, E. (1961). *The Structure of Science*. Harcourt, Brace & World, New York.

Nagel, E. (1998). Issues in the logic of reductive explanations. In Curd, M. and Cover, J., editors, *Philosophy of Science: The Central Issues*, pages 905–921. W. W. Norton & Co., New York.

Nickles, T. (1973). Two concepts of intertheoretic reduction. *Journal of Philosophy*, 70:181–201.

Nickles, T. (1985). Beyond divorce: Current status of the discovery debate. *Philosophy of Science*, 52(2):177–206.

Peskin, M. E. and Schroeder, D. V. (1995). *An Introduction to Quantum Field Theory*. Perseus Books, New York.

Petz, D. (1990). *An Invitation to the Algebra of Canonical Commutation Relations*. Leuven University Press, Leuven.

Popper, K. (1959). *The Logic of Scientific Discovery*. Hutchinson, London.

Post, H. (1971). Correspondence, invariance, and heuristics: In praise of conservative induction. *Studies in the History and Philosophy of Modern Science*, 2(3):213–255.

Primas, H. (1998). Emergence in exact natural sciences. *Acta Polytechnica Scandinavica*, 91:83–98.

Psillos, S. (2001). Is structural realism possible? *Philosophy of Science*, 68(S3):S13–S24.

Radder, H. (1991). Heuristics and the generalized correspondence principle. *British Journal for the Philosophy of Science*, 42(2):195–226.

Reed, M. and Simon, B. (1975). *Methods of Modern Mathematical Physics II: Fourier Analysis, Self-Adjointness*. Academic Press, New York.

Reichenbach, H. (1938). *Experience and Prediction. An Analysis of the Foundations and the Structure of Knowledge*. University of Chicago Press, Chicago.

Rejzner, K. (2016). *Perturbative Algebraic Quantum Field Theory: An Introduction for Mathematicians*. Springer, New York.

Rejzner, K. (2019). Locality and causality in perturbative algebraic quantum field theory. *Journal of Mathematical Physics*, 60(12):122301.

Rieffel, M. (1989). Deformation quantization of Heisenberg manifolds. *Communications in Mathematical Physics*, 122(4):531–562.

Rieffel, M. (1993). *Deformation quantization for actions of $\mathbb{R}^d$*. Memoirs of the American Mathematical Society. American Mathematical Society, Providence, RI.

Rohrlich, F. (1990). There is good physics in theory reduction. *Foundations of Physics*, 20(11):1399–1412.

Romano, D. (2016). Bohmian classical limit in bounded regions. In Felline, L., Ledda, A., Paoli, F., and Rossanese, E., editors, *New Directions in Logic and the Philosophy of Science*, volume 3 of *SILFS*, pages 303–318. London, College Publications.

Rosaler, J. (2015a). "Formal" versus "empirical" approaches to quantum-classical reduction. *Topoi*, 34(2):325–338.

Rosaler, J. (2015b). Is de Broglie–Bohm theory specially equipped to recover classical behavior? *Philosophy of Science*, 82(5):1175–1187.

Rosaler, J. (2016). Interpretation neutrality in the classical domain of quantum theory. *Studies in the History and Philosophy of Modern Physics*, 53:54–72.

Rosaler, J. (2018). Generalized Ehrenfest relations, deformation quantization, and the geometry of inter-model reduction. *Foundations of Physics*, 48(3):355–385.

Rosenstock, S., Barrett, T., and Weatherall, J. (2015). On Einstein algebras and relativistic spacetimes. *Studies in the History and Philosophy of Modern Physics*, 52B:309–316.

Rosenstock, S. and Weatherall, J. (2016). A categorical equivalence between generalized holonomy maps on a connected manifold and principal connections on bundles over that manifold. *Journal of Mathematical Physics*, 57(10):102902.

Rudin, W. (1962). *Fourier Analysis on Groups*. Wiley & Sons, Inc., Hoboken, NJ.

Ruetsche, L. (2002). Interpreting quantum field theory. *Philosophy of Science*, 69(2):348–378.

Ruetsche, L. (2003). A matter of degree: Putting unitary inequivalence to work. *Philosophy of Science*, 70(5):1329–1342.

Ruetsche, L. (2006). Johnny's so long at the ferromagnet. *Philosophy of Science*, 73(5):473–486.

Ruetsche, L. (2011). *Interpreting Quantum Theories*. Oxford University Press, New York.

Sakai, S. (1971). *C*-Algebras and W*-Algebras*. Springer, New York.

Sakurai, J. (1994). *Modern Quantum Mechanics*. Addison-Wesley, New York.

Scheibe, E. (1986). The Comparison of scientific theories. *Interdisciplinary Science Reviews*, 11(2):148–152.

Scheibe, E. (1999). *Die Reduktion physikalischer Theorien*. Springer, Berlin.

Segal, I. (1963). *Mathematical Problems of Relativistic Physics*. American Mathematical Society, Providence, RI.

Slawny, J. (1972). On factor representations and the C*-algebra of canonical commutation relations. *Communications in Mathematical Physics*, 24(2):151–170.

Stanford, P. K. (2003). No refuge for realism: Selective confirmation and the history of science. *Philosophy of Science*, 70(5):913–925.

Stanford, P. K. (2006). *Exceeding Our Grasp*. Oxford University Press, New York.

Steeger, J. and Feintzeig, B. (2021a). Extensions of bundles of C*-algebras. *Reviews in Mathematical Physics*, 33(8):2150025.

Steeger, J. and Feintzeig, B. H. (2021b). Is the classical limit "singular"? *Studies in History and Philosophy of Science Part A*, 88:263–279.

Stein, H. (1981). "Subtler forms of matter" in the period following Maxwell. In Cantor, G. and Hodge, M., editors, *Conceptions of Ether: Studies in the History of Ether Theories 1740–1900*, pages 309–340. Cambridge University Press, Cambridge, UK.

Stein, H. (1987). After the Baltimore Lectures: Some philosophical reflections on the subsequent development of physics. In Achinstein, P. and Kargon, R., editors, *Kelvin's Baltimore Lectures and Modern Theoretical Physics*, pages 375–398. Massachusetts Institute of Technology Press, Cambridge, MA.

Sternberg, S. (1977). Minimal coupling and the symplectic mechanics of a classical particle in the presence of a Yang–Mills field. *Proceedings of the National Academy of the Sciences*, 74(12):5253–4.

Summers, S. (1999). On the Stone–von Neumann uniqueness theorem and its ramifications. In Redei, M. and Stoeltzner, M., editors, *John von-Neumann and the Foundations of Quantum Physics*, pages 135–152. Kluwer, Dordrecht.

Swanson, N. (2019). Deciphering the algebraic CPT theorem. *Studies in History and Philosophy of Science Part B: Studies in History and Philosophy of Modern Physics*, 68:106–125.

Teller, P. (1979). Quantum mechanics and the nature of continuous physical quantities. *Journal of Philosophy*, 76(7):345–361.

Thébault, K. P. Y. (2016). Quantization as a guide to ontic structure. *The British Journal for the Philosophy of Science*, 67(1):89–114.

van Hove, L. (1951). Sur le problème des relations entre les transformations unitaires de la mécanique quantique et les transformations canoniques dela mécanique classique. *Academie Royale de Belgique, Bulletin Classe des Sciences Memoires*, 5(37):610–620.

van Nuland, T. D. H. (2019). Quantization and the resolvent algebra. *Journal of Functional Analysis*, 277(8):2815–2838.

von Neumann, J. (1932). *Mathematical Foundations of Quantum Mechanics*. Princeton University Press, Princeton, NJ. English translation published 1955.

Waldmann, S. (2005). States and representations in deformation quantization. *Reviews in Mathematical Physics*, 17(1):15–75.

Waldmann, S. (2010). Positivity in Rieffel's strict deformation quantization. In *XVIth International Congress on Mathematical Physics, Prague, Czech Republic, August 3–8, 2009. With DVD*, pages 509–513. World Scientific, Hackensack, NJ.

Waldmann, S. (2016). Recent developments in deformation quantization. In Finster, F., Kleiner, J., Röken, C., Tolksdorf, J., editors, *Quantum Mathematical Physics*. Birkhäuser, Cham.

Waldmann, S. (2019). Convergence of star products: From examples to a general framework. *EMS Surveys in Mathematical Sciences*, 6(1):1–31.

Wallace, D. (2001). Emergence of particles from bosonic quantum field theory. Unpublished. arXiv:quant-ph/0112149v1.

Wallace, D. (2006). In defence of naiveté: The conceptual status of Lagrangian quantum field theory. *Synthese*, 151(1):33–80.

Wallace, D. (2011). Taking particle physics seriously: A critique of the algebraic approach to quantum field theory. *Studies in the History and Philosophy of Modern Physics*, 42(2):116–125.

Weatherall, J. (2021). Why not categorical equivalence? In Judit Madarász, Gergely Székely, editors, *Hajnal Andréka and István Németi on Unity of Science*, p. 427–451, Springer Cham.

Weatherall, J. O. (2019a). Part 1: Theoretical equivalence in physics. *Philosophy Compass*, 14(5).

Weatherall, J. O. (2019b). Part 2: Theoretical equivalence in physics. *Philosophy Compass*, 14(5).

Weinless, M. (1969). Existence and uniqueness of the vacuum for linear quantized fields. *Journal of Functional Analysis*, 4(3):350–379.

Weinstein, M. (1978). A universal phase space for particles in Yang–Mills fields. *Letters in Mathematical Physics*, 2:417–20.

Weyl, H. (1950). *The Theory of Groups and Quantum Mechanics*. Dover, Mineola.

Wigner, E. (1959). *Group Theory and Its Application to the Quantum Mechanics of Atomic Spectra*. Academic Press, Cambridge, MA.

Williams, P. (2018). Scientific realism made effective. *British Journal for Philosophy of Science*, 70(1):209–237.

Woodhouse, N. M. J. (1997). *Geometric Quantization*. Oxford University Press, Oxford.

Worrall, J. (1989). Structural realism: The best of both worlds? *Dialectica*, 43(1–2):99–124.

Yaghmaie, A. (2021). Deformation quantization as an appropriate guide to ontic structure. *Synthese*, 198:10793–10815.

Zahar, E. (1983). Logic of discovery or psychology of invention? *British Journal for Philosophy of Science*, 34(3):243–61.

Zalamea, F. (2018). The twofold role of observables in classical and quantum kinematics. *Foundations of Physics*, 48(9):1061–1091.

Zworski, M. (2012). *Semiclassical Analysis*. American Mathematical Society, Providence, RI.

Acknowledgments

This material is based upon work supported by the National Science Foundation (Grant No. 1846560). I am indebted to Jer Steeger and Kade Cichella for many conversations that helped refine the ideas presented here, and to Klaas Landsman for detailed comments on a draft of this work. I would also like to thank Jim Weatherall for helping me see this work through to completion. Finally, I'm grateful for Jess and Raven, without whose support this work would never have been completed.

Cambridge Elements ≡

The Philosophy of Physics

James Owen Weatherall
University of California, Irvine

James Owen Weatherall is Professor of Logic and Philosophy of Science at the University of California, Irvine. He is the author, with Cailin O'Connor, of *The Misinformation Age: How False Beliefs Spread* (Yale, 2019), which was selected as a *New York Times* Editors' Choice and Recommended Reading by *Scientific American*. His previous books were *Void: The Strange Physics of Nothing* (Yale, 2016) and the *New York Times* bestseller *The Physics of Wall Street: A Brief History of Predicting the Unpredictable* (Houghton Mifflin Harcourt, 2013). He has published approximately fifty peer-reviewed research articles in journals in leading physics and philosophy of science journals and has delivered over 100 invited academic talks and public lectures.

About the Series
This Cambridge Elements series provides concise and structured introductions to all the central topics in the philosophy of physics. The Elements in the series are written by distinguished senior scholars and bright junior scholars with relevant expertise, producing balanced, comprehensive coverage of multiple perspectives in the philosophy of physics.

Cambridge Elements ≡

The Philosophy of Physics

Printed in the United States
by Baker & Taylor Publisher Services

Printed in the United States
by Baker & Taylor Publisher Services